ROLE OF BIOTECHNOLOGY IN AGRICULURE

Editors
B.N. PRASAD
G.P.S. GHIMIRE
V.P. AGRAWAL

ISBN 81-204-0720-2 OXFORD & IBH

INTERNATIONAL SCIENCE PUBLISHER
New York
and
OXFORD & IBH PUBLISHING CO. PVT. LTD.
New Delhi   Bombay   Calcutta

ISBN 1-881570-12-6   ISP
ISBN 81-204-0720-2   OXFORD & IBH

Jointly published in U.S.A. by International Science Publisher
28 West 27 Street (second floor), New York, NY 10001, and
in India by Mohan Primlani for Oxford & IBH Publishing Co. Pvt. Ltd.
66 Janpath, New Delhi 110 001. Laser typeset by Indus Publishing Co.
Processed and printed in India by Chaman Enterprises
R-69/1, Ramesh Park, Laxmi Nagar, Delhi 110 092.

Foreword

Agriculture is the main occupation of the people in the SAARC countries. Scientific and technological advancements related to the agricultural fields are, therefore, very valuable to us.

I was happy to learn that the First Regional Conference of the Association of Plant Physiologists of the SAARC Countries (APPSC) was held in Kathmandu in October, 1988. The conference on the ''Role of Plant Physiology and Biotechnology on Crop Productivity'' was a useful one. I hope, therefore, that the proceedings of the conference published in a book form will be useful for the scholars and researchers in our part of the world.

Dr. B.N. Prasad, Secretary-General of the conference, with his team, and the contributors deserve appreciation for their commendable work.

G.P. KOIRALA
Prime Minister, Nepal

Message

It gives me great pleasure to write a few words about the proceedings of the First Regional Conference of the Association of Plant Physiologists of SAARC Countries (APPSC). The Conference on the "Role of Plant Physiology and Biotechnology on Crop Productivity" was held in Kathmandu from October 3-7, 1988. It was a landmark in the history of science and technology. It was attended by 110 national and international scientists and their valuable deliberations were held and important decisions made.

Those are now published in the form of proceedings for the use of researchers and scholars and I am sure that the proceedings of APPSC will be useful, especially for the scientists of Third World countries.

It is an acknowledged fact that agriculture is of prime importance for the countries of South Asia as a great majority of their inhabitants depend on agriculture directly or indirectly.

Dr. B.N. Prasad, Secretary General of the conference together with his team and other eminent contributors deserves appreciation and commendation for this useful venture.

RAM HARI JOSHY
Minister of Education & Culture
His Majesty's Government
Kathmandu, Nepal

Message from Ambassador Kant Kishore Bhargava, SAARC Secretary-General

The publication of the proceedings of the Conference on the ''Role of Plant Physiology and Biotechnology on Crop Productivity'' held in Kathmandu from October 3-7, 1988 is to be greatly welcomed for it will be of direct use and benefit to scientists in the Member States of SAARC. At the recently held fifth SAARC Summit in Male, the SAARC Leaders emphasised the importance of Biotechnology for the long-term food security of developing countries as well as for medicinal purposes and decided that cooperation should be extended to this field and, in particular, to the exchange of expertise in genetic conservation and maintenance of germplasm banks.

I congratulate Dr. B.N. Prasad, Secretary-General of the Conference and the members of his team for the painstaking work which they have undertaken to make available the findings of the Conference to a wider readership in our region.

KANT KISHORE BHARGAVA

Preface

The Association of Plant Physiologists of SAARC Countries (APPSC) was established on December 21, 1987 during the International Conference of Plant Physiology in India. This association of SAARC scientists has been established to facilitate research activities in SAARC countries in the field of Plant Physiology and Biotechnology. The Executive Committee of APPSC held on February 21, 1988 in New Delhi, India decided to hold its "First Regional Conference on the Role of Plant Physiology and Biotechnology on Crop Productivity" in the first week of October 1988, in Kathmandu, Nepal.

This book is based on the research papers presented at the Conference. Dr. S.Ramachandran, Secretary, Department of Biotechnology, Ministry of Science and Technology, Government of India, delivered a Keynote lecture, "Biotechnology and Plant Sciences Opportunities for SAARC Countries" in the inaugural session of the conference.

Eminent scientists and scholars from South Asian countries presented papers which would be very useful and helpful for crop production.

The edited and revised papers included in this publication cover a wide range of research aspects in the field of plant biology and agricultural biotechnology. This book should also be helpful to enhance research activities in SAARC countries. It could be widely used in developing countries by researchers, technologists working in various fields of plant biology and agricultural.

We are thankful to all sponsors specially HMG Ministry of Education and Culture/Ministry of Agriculture, Nepal for providing financial assistance for the publication of these proceedings of the conference and other scientific activities.

EDITORS

Acknowledgements

The success of any conference depends upon the participation from people from various walks of life.

First of all I express my gratitude to Rt. Hon'ble G.P. Koirala, Prime Minister of Nepal for the foreword and good wishes and I am also grateful to Hon'ble Ram Hari Joshy, Minister for Education and Culture, Nepal for the message and good wishes for this publication. I am also thankful to His Excellency K.K. Bhargava, Secretary General, SAARC for the message and good wishes. I express my sincere thanks to Dr. S. Ramachandran, Secretary, Department of Biotechnology, Government of India for delivering the keynote address at the inaugural session of the conference. I am thankful to His Majesty's Government (HMG), Nepal for providing financial assistance for publication of the proceedings of the conference and other scientific activities. I express sincere thanks to all sponsors of this conference who gave us financial as well as organisational help. I am thankful to the President of APPSC, Prof. S.N. Mathur and to Shri Mahesh Kumar Upadhya, ex. Vice Chancellor, T.U., who as the chairman of the Organising Committee made every possible effort to ensure the success of the conference. I am also thankful to the members of Organising committee, specially Dr. Prof. V.P. Agrawal, Director, Research Laboratory for Agricultural Biotechnology and Biochemistry (RLABB) who helped me during the last three years in all respects.

Last, but not the least, I am thankful to Prof. B.C. Malla, Ex. Vice-Chancellor, Tribhuvan University, Chairman of APPSC Regional Conference Trust and its members, Dr. V.P. Agrawal, Dr. G.P.S. Ghimire and Dr. R.B. Basnet who have given their valuable time and suggestions for the publication of these proceedings. I am also thankful to the contributors of research papers, editors and Dr. N. Joshee who have helped us in the screening and editing of the papers for these proceedings.

I believe the book will be helpful and useful for researchers, scientists and technologists in the field of Plant Physiology and Biotechnology in the third world countries.

B.N. Prasad
Secretary-General

Contents

1

Biotechnology and Plant Sciences Opportunities for SAARC Countries

S. Ramachandran and S. Natesh

Department of Biotechnology,
Government of India, New Delhi

INTRODUCTION

The term biotechnology connotes different meanings in various contexts and thus is understood variously in various countries. In advanced countries, such as the USA, Canada, U.K. etc., the term primarily means recombinant DNA technologies. In India, however, the term biotechnology is used in a much broader sense to include the recent dramatic developments in the areas of molecular biology and genetics, techniques in recombinant DNA, cell fusion in plants animals and microbes, animal and plant cell and tissue culture, micromanipulation of embryos and embryo transfer, molecular and cellular immunology, enzyme, organelle and whole cell immobilisation, protein and carbohydrate engineering, and fermentation technology and process engineering, including computerisation. The following broad definition of the word biotechnology is likely to place it in a proper perspective. Biotechnology is the application of biological organisms and molecules to technical and industrial processes. It generally implies the application of novel microbes or other living systems altered or modified by man through various techniques of genetic engineering, cell fusion, etc. Biotechnology is not a new discipline or field of science. It is a set of novel tools and techniques which enables us to manipulate the core of all living matter, i.e., the DNA itself in a manner resulting in the expression of enhanced or even totally new properties that may have never existed before in plants, animals and micro-organisms, thus providing vast opportunities for its application in many areas of health care, energy, environment and industry. The dramatic acceleration in our understanding of the organisation and expression of primitive and highly evolved living systems has been largely due to the availability of these new techniques. The full potential of these techniques has yet to be realised. However, even in the limited period of about a decade, their potential applications have been demonstrated in the most dramatic manner. It is clear that if developed and applied properly, biotechnology can offer solutions to many major national problems and needs. This paper discusses the application of biotechnology to agriculture and allied areas.

The major problems and needs of the SAARC countries also become the major challenges and opportunities for the scientists and technologists of these countries. There are several common problems which need to be tackled.

One of the most pressing problems is that of population. The population of India, for example, has been growing at an alarming—and unacceptably high—rate, and poverty is still widespread. By the turn of the century, the population is expected to cross one billion. To feed this burgeoning multitude while concomitantly promoting ecological sustainability is of the utmost priority. Therefore increasing agricultural production and productivity assume great significance without doubt the yield improvement of major cereal crops in India, such as wheat and rice, is remarkably good, thanks to the green revolution of the sixties and seventies, but corresponding improvement in the production of millets, pulses and oilseed crops has yet to be achieved. The available techniques have not been adequately applied to these crops. The average agricultural productivity is still well below that obtained in several other countries. The country has been importing large quantities of edible oil during the last few years.

The situation is similar with regard to the production and productivity of livestock. Although India has more than 30% of the world's cattle population, the production of useful products is less than 10%. In spite of the fact that the country has a coastline of over 7,000 kms, and over 27,000 kms of rivers not to mention over 100,000 kms of canals and channels, the harvest of fish is just over 2 million tonnes per annum.

Against this background is the prospective plan for the country for 2001 A.D. which requires the production of 240 million tonnes of food grains (against the present 166 million tonnes), 12 million tonnes of fertilisers, 80 million tonnes of milk and milk products, as well as achieving 100% primary health care coverage of the population. Hand in hand, the present trend of degradation of forests has to be arrested—and reversed; the high rate of desertification of land has to be stopped and more than 60 million hectares of degraded land have to be put under green cover.

The situation for other SAARC countries is more or less similar. How can these countries meet these challenges? What opportunities can be taken advantage of in accomplishing the formidable tasks in the relatively narrow time-frame of just over a decade? What are the possibilities for co-operation among these countries?

AGRICULTURAL BIOTECHNOLOGY

Doubtless, the task is formidable and challenging. Obviously, in the agricultural sector it is absolutely essential that the production and productivity of the major food crops be pushed up. The yields of major cereal crops have now reached a plateau; the conventional techniques have been well exploited, and any further yield gains are unlikely without using newer approaches. As for the yields of millets, pulses and oilseed crops, there is still scope for yield improvement, but given the narrow time-frame, it would be very essential to employ the recently developed biotechnology tools in addition to the conventional ones. Clearly, strategies for all these crops will have to be in terms of :

 i) Improving the quality and quantity proteins;
 ii) Minimising the pre- and post-harvest damage caused by pests, pathogens and environmental stresses through the development of stress-tolerant types, and biological pest control;
 iii) Increasing the fertility of soil and improving the availability of nutrients through biological nitrogen fixation (both symbiotic and non-symbiotic) and use of mycorrhizal organisms;
 iv) Mass multiplication of elites and high-yielding types and generation of more variation through tissue culture;
 v) Rapid *in vitro* multiplication of specific plant species for biomass and green cover.

Over the last 25 years there has been a remarkable increase in our knowledge of the factors governing the quality and quantity of cereal grain proteins. In recent years, techniques of molecular biology have been applied to gain new information. Newer discoveries such as the existence of multigene families for storage proteins such as zein and hordein, cloning of seed protein genes, and r-DNA technology are rapidly widening the horizons of our knowledge regarding the organisation, expression and regulation of seed protein genes. Scope now exists for altering the quantity and quality of storage proteins through genetic engineering. For example, the bean seed protein phaseolin has been transferred to sunflower through Ti-plasmid vectors. It may thus be possible to engineer seeds that produce surplus proteins, or to alter the quality of such proteins by site-directed mutagenesis. It may also be possible to integrate genes of specific animal proteins in place of storage protein genes.

Stress-tolerant Crops

One of the constraints to yield improvements is the presence of stress conditions—both abiotic and biotic—during the growth of crops. The abiotic stresses include anaerobiosis, moisture, temperature, salinity and so on. The biotic stresses could be various pathogens and pests.

Agriculture in India is still largely dependent on the vagaries of the monsoon. The yields are relatively proportional to the extent of the monsoon, thus seriously affecting the country's economy. Although major attention is being paid to evolving a more integrated system of water and soil management, it is imperative that our farming be made less dependent on water and increasingly utilise for agricultural purposes marginal and less productive land. Most of India's millets, pulses and oilseeds are grown under rainfed and suboptimal conditions. Here, the rapid mass screening of segregating populations under controlled environment using tissue explants or cell lines for stress situations, such as drought, salinity and temperature, could greatly reduce the time required for selection of mutant lines. Similarly, somaclonal variation has been shown to be a method of high potential for generating phenotypic variation. Recently, genetic engineering techniques have been employed by Advanced Genetic Sciences, a biotechnology company, to provide frost tolerance to strawberry, apple, and potato plants. In this study, bacteria of the genus *Pseudomonas* were genetically altered by removal of the gene for ice nucleation protein and applying it to the plant surface.

One of the oft-cited examples is the use of genetic engineering to produce elite crop varieties with novel herbicide resistance so that when the herbicide is applied to the field, only the weeds are eliminated. Single genes from microorganisms have been cloned and expressed in plants to confer the desired resistance (e.g. against glyphosate, roundup, etc.).

Among the biotic stresses, pathogens of various nature—bacteria, viruses, fungi, etc.—cause great damage to standing crops. Here, again, rapid large-scale cell screening *in vitro* in the presence of the pathogen or the toxin produced by it could be very useful. Somaclones of crops resistant to *Phytophthora* and *Alternaria* have been isolated. It may be possible to obtain Tungro virus and leaf-hopper resistance in rice, fungal-resistant types in groundnut, pigeon-pea, sugar-cane etc. Genetic engineering techniques have been utilised through the expression of viral coat protein for induction of tobacco mosaic virus resistance in tomatoes. Transgenic tobacco plants resisting the cucumber mosaic virus have recently been reported. Similarly, expression of viral coat proteins of the alfalfa mosaic virus (AlMV), and potato virus (PVX) in transgenic plants results in protection of these from infection by AlMV and PVX. In the present context, viral diseases of potato, pigeon-pea, mung bean etc. could be tackled through this technique.

Another major contribution of biotechnology would be in providing improved and highly sensitive methods for selection of pathogen-free planting material, breeding stocks and segregants using ELISA monoclonal antibodies and C-DNA probes. It would be very useful to develop such probes for the

Alternaria disease of *Brassica*, the blast disease of rice, the *Ascochyta* disease of chick-pea, among others.

Pests, particularly insects cause maximum damage to both standing crops and stored produce. The damage caused to important crops such as chick-pea, pigeon-pea and cotton can bring down yields by 25% to 40%. Control of pests in the past has mainly been achieved through the application of agrochemicals. However, chemical control is expensive and environmentally undesirable as the chemicals are not target-specific and often leave toxic residues. Considerable effort is being focussed on improving the crop's own defences against attack. Using Ti-plasmid of *Agrobacterium*, BT-toxin genes have been transferred to tobacco and tomato plants. The newly acquired gene is stable and is transmitted in Mendelian fashion; it protects the transgenic plants from damage caused by larvae of the tobacco hornworm. Limited-scale field trials of the engineered plants has been successfully carried out. The use of BT-genes to confer resistance to insects have been widely welcomed, and is likely to be very useful in a variety of crop plants (e.g., agasint the gram pod borer in pigeon-pea and chick-pea, against the red cotton bug and bollworm in cotton, etc.) However, the desirability of the toxin in the edible portion of the plant has been questioned, although BT-toxins have long been considered 'safe' pesticides.

An alternative approach could be to transfer resistance through genetically altered bacteria, such as *Pseudomonas*, which are surface colonisers. An attempt in this direction has been made by the Monsanto group. The delta endotoxin from *Bacillus thuringiensis* has been cloned into the corn-root-colonising bacterium *Pseudomonas flourescens*. The advantages would be increased stability (in the field, the toxin is inactivated by sunlight and persists for only a few days) and decreased horizontal transmission to other soil bacteria. The application to field test the recombinant organism is still pending with the EPA.

Biological control of insect pests has been tried using a variety of micro-organisms—bacteria, viruses, fungi, etc. Some of these have already been commercially exploited. More studies are required in this direction.

Soil Fertility and Nutrient Availability

Certain micro-organisms associated with the plant rhizosphere have beneficial effects on plant growth. One of these is the ability to fix atmospheric nitrogen directly and to make it available to the crop. The micro-organisms may be free-living (e.g., the bacteria *Klebsiella* and blue-green algae), or associated with the roots of either leguminous plants (e.g., the bacterium *Rhizobium*), other dicotyledons (e.g., *Frankia*, an actinomycete) or monocots such as grasses, wheat and maize (e.g., *Azospirillum*). It is now well established that the effect of nitrogen fixation is most important for leguminous plants. The extent of nitrogen fixed varies and depends on a number of factors, but strains of rhizobial species are available which can add 60-65 kgs of nitrogen per hectare. Among the non-symbiotic nitrogen fixers, the combination of *Azolla* and blue-green algae (BGA) when incorporated in wetland rice fields, can make a substantial contribution to the nitrogen pool. Wider use of rhizobia and BGA for our pulses, oilseed legumes and wetland crops would add the equivalent of 3-5 million tonnes of urea per annum at a cost which is a small fraction of that required for chemical fertilisers. Thus such micro-organisms have considerable potential as biofertilisers.

The beneficial effect of *Azospirillum* may derive not so much from its nitrogen-fixing abilities, as its stimulatory effect on root development, probably caused by secretion of growth hormone-like substances. Many soil microorganisms, in addition, may enhance the mineral uptake of the plant, for example, the solubilisation of phosphate from the soil.

In Vitro Multiplication of Elites and High-yielding Types

The technique of plant tissue culture has proved to be a very handy tool, in that it can be a means of multiplying plants at remarkably fast rates under controlled phytosanitary conditions. This would be particularly useful for clonal propagation of perennial crops which are normally outbreeders and reproduce only through seeds. Coconut and oil palm, for example, are ideal candidates for this technique. In India, most of the traditionally grown coconut have a low productivity at 30-50 nuts per tree per year. However, there are elites and hybrids which produce 3-5 times this number. *In vitro* clonal propagation of such trees in large numbers could substantially increase productivity of this important oilseed crop. Efforts in this direction are underway. Similarly, oil palm, which has the highest productivity among all the edible oil-bearing crops at 4-6 tonnes of oil per ha per year is another species that is being regenerated through *in vitro* techniques. A major programme for taking up oil palm cultivation on a large scale is underway. Another example of clonal propagation is cardamom. Through micropropagation, elites of cardamom can be multiplied at significantly high rates.

Tissue culture of arborescent species is another area of great importance. Especially for enhancement of green cover, and production of biomass, rapid multiplication of woody species is a highly desirable goal. Propagation of proven genotype of a tree necessitates the use of 'mature-tree' explants. However, the regeneration of trees from adult individuals is not easy as there is some 'maturity barrier' associated. Although several tree species have been regenerated *in vitro* using explants derived from seedlings, examples of successful culturing of older plant materials are not many. These include sequoia, white spruce and eucalyptus. In India, mature teak, eucalyptus, sandalwood trees and bamboo, among several others, have been regenerated *in vitro*. Elite trees or 'super trees' of teak, *Pinus*, *Bombax* (silk cotton tree), and *Eucalyptus* hybrids can be selectively propagated so that their clones become available for planting in large numbers.

Germplasm Banks for Plants

The Indian subcontinent has been the origin of a large number of culti-vated varieties of plants and their wild relatives and many of these have been shown to have several desirable genetic traits. The common technique for the preservation of these is through the seeds. However, for several species of agrihorticultural importance, the conventional methods are inadequate or inap-propriate, particularly if the species happens to be vegetatively propagated, recalcitrant, etc. As an adjunct to the conventional methods, preservation of *in vitro*-raised tissues of such species could be carried out under cryogenic con-ditions, both on a short- and long-term basis. A National Plant Tissue Culture Repository has been established at New Delhi for this purpose. Perhaps this is the first ever attempted at the national level anywhere in the world.

ANIMAL BIOTECHNOLOGY

In the largely agro-based economy of the SAARC countries, livestock are still the mainstay for the livelihood of the millions in the villages. Animals, espe-cially cattle and buffalo, are used for a variety of agricultural practices. The economic importance of cattle, buffalo, sheep, goat, swine, poultry, etc. for the production of milk, egg, meat and leather cannot be overemphasised. India has an estimated population of about 200 million cattle, nearly half this number of buffalows, 82 million goats, 41 million sheep and about 10 million pigs. These animals compete along with humans for space, water and the scarce feed and fodder. The country presently produces about 40 million tonnes of milk per annum and our target for 2000 A.D. is 80 million tonnes of milk. The average milk production in the country is a mere 400 litres per cow and 800 litres per buffalo per lactation. At the present rate of yield, reaching a target of 80 million tonnes in the normal course would mean almost doubling the animal population. Such a course would put tremendous pressure on the already scarce resources for sustaining a huge population of low-productivity animals. It is rather impractical and unacceptable.

However, a small precentage of indigenous breeds are known to be high yielders, producing over 3000 litres of milk per lactation. An estimated 60,000 MW equivalent of draught-animal power (equivalent to over 100 million bul-locks) will be required in the country by 2000 A.D. As in the milch animals, there are superior draught-animal breeds also available in the country.

It has long been recognised that herd improvement through selection and multiplication is essential to achieve higher productivity. Therefore, genetic improvement of indigenous breeds was taken up in India in a major way for several years. However, the conventional methods of herd improvement through cross-breeding and artificial insemination failed to produce herd improvement as fast as required because of the low female reproductive rate and the consequent low selection intensities: a cow in its productive life can only produce 6 or 7 calves, whereas through the use of multiple ovulation and the embryo transfer (MOET) technique the potential could be raised to a calf production of 50-60 in a single cow's productive life.

For large-scale multiplication of a superior animal, MOET is the method of choice. The potential of MOET for herd improvement can be further enhanced by some of the associated technologies, such as embryo splitting, embryo sexing, embryo cryopreservation, *in vitro* oocyte maturation, *in vitro* fertilisation, etc. By the judicious application of MOET and associated technologies, it is possible to double or even treble the productivity in less than 15 years without increasing the existing total population of animals.

The administration of growth hormones has been shown to increase productivity in animals significantly. A 30% increase in productivity is reported from Indian studies. It is also possible to develop transgenic animals with desirable traits introduced through micromanipulation.

A focussed and comprehensive herd improvement programme at the national level as an S & T project in mission mode entitled 'Herd Improvement for Increased Productivity using Embryo Transfer Technology' was launched in India recently. This project aims at introduction of ET in Indian cows and buffaloes through the creation of an extensive infrastructure, skill pool of highly trained personnel and also a seed stock of superior germplasm as an embryo bank. Many premier institutions are participants in this programme. Under this project, several ET cow calves have already been born. Non-surgical ET has been successfully achieved in several buffaloes. Unlike in cows, ET is perhaps the only method for herd improvement in buffaloes as import of germplasm from other countries is not possible in this case since all the known exotic breeds of buffaloes exist in India. The project also achieved major success in cryopreservation and embryo-splitting. Towards the completion of the present project phase of 5 years in 1992, it is expected that a capability to produce a seed stock of 10,000 exotic embryos per year will be generated.

A network of infrastructural facilities including an ET main laboratory at Bidaj, three other R & D laboratories (at National Dairy Research Institute, Karnal; Indian Veterinary Research Institute, Izatnagar; and National Institute of Immunology, New Delhi), 4 ET Regional Centres and 25 ET State Centres are being set up under this project. The ET project activities will have strong linkages with the ongoing herd improvement programmes, especially the extensive artificial insemination network. Any concerted attempt to quickly improve the productivity of cattle and buffaloes cannot be achieved without sterilising the low potential males of the species that roam and breed freely in the countryside. A new single injection male sterilising agent called 'Talsur' has been developed for field use in conjunction with ET and AI programmes. Once the technical capability for ET in cows and buffaloes is fully achieved, it should be possible to extend it to other economically important animals, such as sheep, goat, swine, camel, etc.

Selection and multiplication of superior breeds alone cannot make the herd improvement programme a success. Curative and preventive aspects of animal disease management are important elements in achieveing higher productivity. The modern biotechnological tools of genetic engineering, cell cultures, cell fusion, molecular biology, protein engineering, etc. have facilitated the development of effective, safe and also cost-effective vaccines and highly sensitive immunodiagnostics against several animal diseases. Several important vaccines are being produced in India.

AQUACULTURE BIOTECHNOLOGY

Even though reliance on fish as an important source of food has a long history, the science of aquaculture is relatively young. Our understanding of the breeding reporduction and genetics of fish is not adequate. Modern biotechnologies in aquaculture offer vast potentials through hormonal and other mechanisms. The reproductive cycle and the growth rate can be manipulated, resulting in significant increases in fish productivity. Even the manipulation of the chromosome sets in fish to obtain triploid and tetraploid species results in sterile fish, which ensures faster growth as no energy is wasted in reproduction. Already several species of fish of economic importance have been successfully microinjected in the early embryonic stage with the growth hormone gene, resulting in species achieving a higher growth rate and even size. The procedure of single eye ablation in prawns to uniformly regulate optimal spawning through controlling the levels of the hormone hCG is now a common practice. The development of high-yielding varieties of fish for growth in shallow and temporary ponds and stagnant water areas such as paddy fields would enormously enhance the proteinous food available as well as the income of rural communities.

POULTRY BIOTECHNOLOGY

Biotechnology has major applications in modernished popultry production. Substantial progress has been achieved in the scientific management of poultry for layers and broilers in large vertically integrated programmes involving properly identified parent breeder stock, automatic hatchery operations, feed mill and processing and packaging facilities. Research carried out on the use of dwarf varieties consuming less feed with better feed conversion ratio has demonstrated their suitability in poultry production programmes, Further, by genetic manipulation, the desired genes may be introduced into other varieties for better efficiency. There is much scope for the application of molecular biology in identifying and converting non-conventional sources of food and in the use of hormones for hastening the growth rate. It has also been demonstrated that identifying the genes responsible for certain diseases, such as the Leucosis Complex, and deleting such genes while developing newer varieties would help in producing disease-resistant flocks. The SPF (Specific Pathogen Free) concept has made major contributions to the poultry industry, particularly in the production of vaccines by avoiding pathogens which are vertically transmitted. Better feed formulation through computers is yet another breakthrough. These innovations have already shown promising results in developing immunodiagnostic methods in poultry husbandry.

OPPORTUNITIES FOR CLOSER CO-OPERATION

There is a major thrust on agricultural biotechnology the world over. According to some estimates, agri-business will corner over 50% of the world market in the coming decade. Several transnational companies in the USA, Europe and Japan have already invested huge sums of money in this area. However, in India, as in most other developing countries, R & D efforts so far have been mainly through governmental funding. In recognition of the tremendous importance of biotechnology, the Government of India set up the National Biotechnology Board in 1982 and four years later created a full-fledged Department of Biotechnology, to act as a focal point for the planning, promotion and co-ordination of biotechnology programmes in the country. The major programmes initiated are: integrated manpower development in biotechnology, creation of infrastructural facilities and support to R & D programmes. Under the integrated manpower development programme, several programmes have been initiated. These include the post-graduate and post-doctoral course, short-term training programme, biotechnology associateships and the Visiting Scientists from Abroad Programme.

Seventeen universities/other institutions have been strengthened to offer two-year post-graduate course (M.Sc./M.Tech) in biotechnology. About 10-20 students per year are trained in each of these institutions. In the area of agriculture and allied sciences, the following Universities offer courses (2-year programmes):

Name of the Institution	Course Offered
1. Indian Agricultural Research Institute, New Delhi	M.Sc. in Molecular Biology and Biotechnology
2. G.B. Pant University of Agriculture & Technology, Pantnagar.	M.Sc. (Agric.) Biotechnology
3. Tamil Nadu Agricultural University, Coimbatore	M.Sc. (Agric.) Biotechnology

4.	Assam Agricultural University Jorhat.	M.Sc. (Agric.) Biotechnology
5.	Indian Veterinary Research Institute, Izatnagar	M.Sc./M.V.Sc. (Vet.) Biotechnology
6.	University of Goa, Panaji	M.Sc. in Marine Biotechnology

Each of these takes about 10 to 12 students per year. It is possible that one seat could be reserved for a participant from SAARC countries other than India.

The short-term training courses have been launched. High-tech based specialised areas as well as expertise and infrastructure for conducting these courses has been identified by the Government. The objectives of this programme are to re-orient the existing manpower within a short period to meet the immediate requirement. Mid-career scientists having permanent facilities in research laboratories/universities from all over the country are sponsored by their host institute for training in a field relevant to the requirement of the parent institute. For this, two or three guest faculty from within the country and one foreign expert, if expertise is not indigenously available in India, are invited. The duration of each course is 2-4 weeks and about 12-18 participants are trained in each workshop. The Department of Biotechnology organised 23 such courses in 1984-86 and in these about 360 persons were trained. During 1986-87, 17 such courses were conducted. During 1987-88, 15 courses were organised.

Most of these courses are technique-oriented. The major topics covered in agriculture and allied areas in these courses include the following:

(i)   Plant Tissue and Cell Culture—methods and applications
(ii)   Laboratory animal science
(iii)   Micropropagation of wood species
(iv)   Protoplast Technology
(v)   Advanced training in Plant Molecular Biology
(vi)   Rhizobium Genetics
(vii)   Structure, function and genetics of chloroplast
(viii) Embryo transfer in cattle
(ix)   Elisa in diagnostics
(x)   Organisation and cloning of plant genes
(xi)   Plant biotechnology applications in forestry

It is also possible to include a scientist from among the SAARC countries, as a participant in these training courses.

The visiting Scientists from Abroad Programme is yet another programme through which outstanding scientists are invited to visit for 3 to 6 months in Indian Laboratories.

Besides these, Biotechnology Associateships are awarded to young and senior scientists. All these programmes are to produce an adequate manpower.

Under the programme for Infrastructural Facilities, several national facilities have been established to provide support systems for building up a strong base for R & D and manufacturing. Of relevance here are the germplasm banks, particularly for algal collections, microbial type culture collection, plant tissue culture repository, and animal cell and tissue culture facility. The germplasm banks will collect, acquire, maintain and supply cultures of organisms of relevance to the country. The emphasis is on strong R & D and training programmes. The Department has also strengthened and expanded animal house facilities to maintain and supply disease-free, and genetically as well as microbiologically refined laboratory animals. In addition, a centralised facility for production, import, and distribution of fine chemicals and enzymes, etc. has also been established, mainly to help the scientists engaged in research work. A bioinformatics network has been established to link up specialised centres within the country and to open up international gateways in the rapidly evolving area of biotechnology. Perhaps it may be possible for scientists from other countries to be trained in some of these areas.

The Department has adopted a two-way approach to solving the major problems. While on the one hand major projects are being encouraged to develop and standardise modern techniques in important areas (e.g., biological nitrogen fixation), on the other clear priorities are being identified in terms of agriculturally important crops and specific problems related to them with an integrated approach, within a defined time-frame of 5-8 years. Four major crops—*Brassica*, rice, chick-pea and wheat—have been identified for immediate intensive work. Others will follow.

With all the major initiatives already underway, the coming years should produce exciting results on the Indian agricultural horizon. The Government of India (Department of Biotechnology) would be happy to consider specific proposals from countries in the SAARC region regarding the possibilities of sending scientific personnel for some of the training programmes described above or for any assistance from the infrastructural facilities established in India. I would like to wish a very fruitful co-operation amongst all of us which would not only strengthen our efforts towards finding solutions to our problems, but would increase our capabilities by way of self-reliance.

2

Gene Cloning for Crop Improvement

B. B. Biswas

Bose Institute, Calcutta, India

ABSTRACT

The ability to introduce foreign genes into a wide variety of plants has created a revolution in plant biology. Genetically engineered crop plants, resistant to herbicides, insects, fungi and viruses are appearing in the scene but the progress of the work is slow due to difficulty in transformation and regeneration of most of the important crop plants such as cereals and legumes. Virus resistance alone could provide significant yield protection in important crop plants. The second aspect which has attracted the attention is the development of plants, resistant to different environmental stresses such as, water, salt, high or low temperature, etc. However, the specific genes which can confer resistance to different stresses are not known. Attempts have also been made to imrpove the quality of storage proteins as well as to produce certain pharmaceutical products from transgenic plants. Genetic engineering has provided a large number of vector constructs and regulatory sequences that might allow for accurate targeting as well as gene expression in specific tissues. Our laboratory has been engaged in producing transgenic mung bean plants (*Vigna radiata*) enriched in storage proteins containing methionine and cysteine as well as resistant to yellow mosaic viruses and insects. Finally, the overall picture in designing plants with improved ability to withstand diseases and unfavourable environments by application of genetic engineering is encouraging. However, there is a vast amount of basic work particularly with the crop plants yet to be done and this aspect should be given primary consideration.

INTRODUCTION

Genetic engineering of plants is rapidly becoming a reality and plant gene transfer has become a fertile field now. Gene transfer into dicot plants can generate varieties with decreased herbicide susceptibility and can also be used to study plant gene control. Genetic manipulation of the monocot plants involves certain difficulties. However, these difficulties are now being resolved at a quicker pace by developing such methods as electroporation for the introduction of new genes into monocot plant cells through a tailored vector. Nevertheless, the long-standing inability to regenerate whole plants from transformable cells still remains as an obstacle to the successful genetic engineering in monocots and legumes. A model genetic engineering of a plant consists of the following general steps: (1) Selection of a plant gene whose introduction in other plants would be of positive agricultural value; (2) identification and isolation of such genes; (3) Transference of isolated genes to the plant cell; and (4) Regeneration of complete plants from transformed cells/tissues. It is also to be considered to insert a bacterial/ fungal/viral gene (useful to plants) to plant which will impart a useful characteristic including disease resistance. Gene manipulation techniques coupled with classical breeding programme are likely to result in great improvements in crop production. Successful first steps towards introduction of disease, herbicide and pesticide resistance in plants have already been reported from several laboratories, following the strategies mentioned thus far. The overall picture in designing plants with improved abilities to withstand environmental and biological stresses are encouraging. It is likely that more genetically engineered crop plants with improved quality of protein and oil will appear in the scene rapidly. However, there is a vast amount of work particularly basic in nature still to be done which should be given prime consideration. Our efforts are at present directed towards this objective.

INTRODUCTION OF FOREIGN GENES TO PLANT

(a) Ti-plasmid used as the vector (derived from *Agrobacterium tumefaciens*): The concept of two types of vector has already emerged: (i) Integrative vectors whereby advantage is taken of the fact that the tumour-inducing genes are not required for the infection and that any DNA sequence that is inserted between the two borders is to be transmitted to the plant. In these vectors, the tumour genes of T-DNA are replaced by sequences of the *E.coli* vector pBR 322. The foreign DNA which is to be integrated into the plant genome is cloned in a pBR vector and then inserted by homologous recombination, via these pBR sequences, into the Ti-plasmid (containing the vir gene and other markers); (ii) Binary vectors whereby the vir region and the borders can be localised on two separate plasmids. There are many constructions already available [1].

(b) Cauliflower mosaic virus used as the vector: The use of CaMV as a vector has certain limitations. Due to the genomic structure of this virus, only a small sequence of DNA (up to 500 bp) can be integrated into the virus genome. The narrow host range (only the *Brassica* family) prevents its use as a universal plant vector. The integration of the DNA through this vector is not stable. However, 35S region of this CaMV genome has a strong promotor activity which is at present being widely used in reconstructing the plant vector (pMON9749 and several other such vectors containing chimeric genes). New selectable marker genes as well as new and more powerful reporter enzymes have already been developed. The most useful selectable marker genes besides NPTII (neomycin phosphotransferase) appear to be hygromycin phosphotransferase (HPT), mouse methotrexate resistant dihydrofolate reductase, the bleomycin resistance gene from Tn5, spectinomycin resistance gene from Tn7 and β-glucuronidase gene (GUS) from *E.coli*, the luciferase from fireflies and the phosphinotricine acetyltransferase, all of which are presently being used by different groups of workers [2]. Stable transformation of plant cells by particle acceleration through electric discharge to introduce DNA-coated gold or tungsten particles into meristems has also been achieved in limited cases. Use of the gemini virus as a vector has also been worked out but no major success achieved as yet.

EXPRESSION AND GENETIC STABILITY OF INTRODUCED GENES

Several genes have been introduced into plant genomes using Ti-plasmid or other derived vectors. So far β-globin of rabbit, alcohol dehydrogenase of yeast and actin and ovalbumin of chicken have been tested in tobacco but none of these genes are properly expressed. On the other hand, the plant gene coding β-phaseolin, a bean storage protein is properly expressed and processed in sunflower cells. When introduced into tobacco, this gene is developmentally regulated and is properly expressed only in the seeds [3]. The light inducible gene for the small subunit of ribulose 1,5-bisphosphate carboxylase of pea retains its potential to be induced by light when it is introduced into petunia cells that are green in tissue culture. Several chimeric gene constructs have been introduced into plant cells. Most of them are composed of the regulatory control sequences of the nopaline synthase gene of the Ti-plasmid and the protein-coding part of the bacterial genes coding for kanamycin resistance. The promotor and poly-A addition site of CaMV gene VI and the promotor of the rubisco gene have also been used for chimeric gene constructions. The antibiotic resistance traits allow the specific selection of transformed cells. In the same way, the introduction of an herbicide-resistant gene into plants has been reported. Foreign genes, integrated into the plant genome, are sexually transmitted to progeny plants, without affecting their expression. Genetic analysis has shown that foreign genes segregate in a Mendelian fashion. T-DNA analysis has demonstrated that transformed seedlings contain a T-DNA stretch identical to that of the parental plants.

APPROACHES TO MAKE PLANTS DISEASE RESISTANT

The resistance of plants to potentially pathogenic micro-organisms is based on multiple biochemical factors. Particularly important are the phytoalexins, absent in healthy plants but which accumulate in response to microbial infection [4,5]. In addition, plants may possess proteins with antibiotic activity [6]. Plant lectins that bind to fungal cell walls are thought to have an antifungal role[7]. It has recently been reported that plant chitinases but not chitin-binding lectins are important antifungal proteins in plants [8]. If chitinase activity could be induced into a plant by ethylene treatment or some other means, the plant might prove resistant to *Trichoderma viride* and other related fungi. Alternatively, chitinase genes could be introduced into the plants which are susceptible to those fungal diseases.

Infection of tobacco plants with TMV results in an increase in the activities of several enzymes and induces the *de novo* synthesis of about 10 proteins that are protease resistant. These proteins accumulate in the intercellular leaf space [9]. The appearance of pathogenesis related proteins (PR) is closely related to the phenomenon of 'systemic acquired resistance' and it has been suggested that such proteins have an antiviral function [10]. It has also been suggested that the ethylene production evoked by TMV infection, is responsible for the induction of PR proteins and several enzymes. Cross protection, inoculation of plants with mild strains of viruses, is often used to reduce losses due to more virulent strains of the tobacco mosaic virus. A chimeric gene containing a cloned cDNA of the virus coat protein under the control of the cauliflower mosaic virus 35S transcript promotor has been introduced into tobacco cells on a tumour-inducing plasmid of *A. tumifaciens*.

The transfected plants produced high levels of the virus coat protein [11] Protection from the virus, manifested as a marked delay in the appearance of symptoms, is dosage dependent and can be partially overcome by inoculation with naked viral RNA. The number of primary sites of infection may be rescued when the virus is encapsulated by the coat protein, and the limited amount of protection to naked viral RNA suggests that there is a second mechanism affecting viral replication or spread. It would be interesting to ascertain whether any of the PR proteins could prevent this spread. Induction of protease inhibitors, which are proteinous in nature, has also been observed in tomato and potato after infection with fungus or after wounding. These proteinase inhibitors might help to make the plant resistant to the attack of pathogens.

ENGINEERING HERBICIDE AND PESTICIDE RESISTANCE

It is relatively easy to produce herbicides selective for plants. Atrazine and diuron interfere with photosynthesis and glyphosate, the sulphonyl ureas and imidazolinones block the synthesis of essential amino acids. But crop plants share these processes with weeds, so crop plants must be protected through differential uptake and metabolism of the herbicide application. What has been found is that atrazine is an effective herbicide for use with maize, which can detoxify the compound but soybean is sensitive. Soybean can be made atrazine resistant by genetic engineering since atrazine can bind with the chloroplast 32K protein.

The other aspect to be considered is to insert a bacterial gene (useful to the plant) into the plant cell, which would impart to the plant a useful characteristic. A gene of the enzyme 5-enolpyruvyl-shikimate-3-phosphate synthase (EPSP synthase) involved in aromatic amino acid synthesis is the target for glyphosphate in *Salmonella typhimurum*. Some of the bacteria contain the gene coding for the resistant form of EPSP synthase (Aro A gene). The mutant form of the enzyme differs from the wild type by a single amino acid. This mutated gene, when transferred into the tobacco plant, induced glyphosate resistance in the plant [12].

There is considerable interest in the biological control of insect pests. The best known example of the use of an organism for biological control of insects is *Bacillus thuringiensis*. This bacterium produces a toxic protein which is active in the insect gut and kills many types of Lepidoptera. The approach is to isolate the gene which codes for toxin from the *Bacillus* and to clone this into crop plants. Encouraging results have already been achieved in the case of tobacco plants. Another approach is to transfer genes for chemicals which control insect behaviour. The marsh pepper produces a substance called polygodial, which when sprayed onto crops prevents aphids from feeding. Polygodial is synthesised enzymatically from farnesylpyrophosphate which is present in all plants. If the genes coding for the enzymes involved could be isolated and transferred to crop plants, it should be possible to produce plants resistant to aphids [13].

The approach currently being followed in Bose Institute is completely different. Tubulin is apparently a well conserved protein in eukaryote, involved in cell division and various other functions. Plant tubulin or fungal tubulin has been found to differ from animal tubulin both immunologically as well as in binding with several drugs, herbicides and fungicides. If such tubulin gene imparting resistance to those chemicals could be introduced into crop plants, it is expected that the plants would be made resistant to those herbicides and fungicides. Another aspect that interests our laboratory is the genetic make up and cloning of several cDNAs and genes from *Vigna radiata*. The isolation and characterisation of cDNA for α- and β-tubulin from mung bean have already been reported [14]. Furthermore, a repetitive DNA of 420 bp has likewise been characterised and sequenced [15] and those sequences/cDNA can be used as markers to analyse the genome. Attempts have also been made to characterise the storage protein from the same plant [16] and cDNA has been cloned. Our

18   *S.S. Biswas*

laboratory has also been successful in transforming and regenerating mung bean plants from explants and a transgenic mung bean plant with respect to some antibiotic resistance has been obtained for the first time using this crop plant as a model system (in press). The future programme to improve upon the storage protein through increasing methionine residues by the insertion of a 2S-gene from Brazil nut, is presently in progress; this 2S storage protein gene is rich in methionine residues.

ANTISENSE RNA

This technique is based on blocking the informational flow from DNA via RNA to protein, by the introduction of an RNA sequence complementary to the sequence of the target, mRNA. Thus an RNA duplex is formed between mRNA and antisense RNA. This duplex RNA is either rapidly degraded or the mRNA is impaired in nuclear processing or it is blocked for translation. Studies done to date indicate that the expression of both sense and antisense α–amylase transcripts is developmentally regulated and both RNAs are present in about an equal amount in barley aleuron tissue. The results obtained thus far with antisense Rubisco SS gene show that it can inhibit plant gene expression using the CaMV-35S promotor [17].

The overall picture in designing plants with improved abilities to withstand environmental and biological stresses is encouraging. It is likely that more genetically engineered crop plants with improved quality of protein and oil will soon appear on the scene. However, there is a vast amount of rather basic work still to be done which must be given primary consideration. The efforts of Bose Institute are presently directed towards this objective.

LITERATURE CITED

1. Zambryski, P. 1988. *Ann. Rev. Genet.*, 22:1-30.
2. Weising, K., J. Schell and G. Kahl. 1988. *Ann. Rev. Genet.*, 22: 421-477.
3. Sengupta-Gopalan, C., N.A. Reichert, R.T. Barker, T.C. Hall and J.D. Kump. 1985. *Proc. Natl. Acad. Sci. US.*, 82:3320-3324
4. Bell, A.A. 1981. *Ann. Rev. Plant Physiol.*, 32: 21-81.
5. Darvill, A.G. and P. Albersheim. 1984. *Ann. Rev. Plant Physiol.*, 35: 243-275.
6. Mirelman, D., E. Galun, N. Sharon and R. Lofan. 1975. *Nature*, 256: 414-416.
7. Roberts, W.K. and C.P. Selitrennikoff. 1986. *Biochem. Biophys. Acta.*, 880: 161-170.
8. Schlumbaum, A., F. Much, V. Vogeli and T. Bollen. 1986. *Nature*, 324: 365-367.
9. Van Loan, L.C. 1985 *Plant Mol. Biol.*, 4: 111-116.
10. Van Loan, L.C. 1975. *Virology*, 67: 566-575.
11. Abel, P.P., N.S. Richard, D. Barun, N. Hoffmann, G.R. Stephen, R.T. Fraley and B.N. Beachy. 1986. *Science*, 232: 738-743.
12. Comai, L., D. Facciotti, W.R. Hiatt, G. Thompson, R.E. Rose and D.M. Stalker. 1985. *Nature.*, 317: 741-744.
13. Dean Peter, D.G. 1975. *Proc. Biotech.*, 1: 301-309.

14. Raha, D., K. Sen and B.B. Biswas. 1987. *Plant Mol. Biol.*, 9: 565-570.
15. Roy, P., N. Bhattacharyya and B.B. Biswas. 1988. *Gene,* 73: 57-66.
16. Bhattacharyya, S.P. and B.B. Biswas. 1990. *Biochem. Intl.*, 21: 667-675.
17. Vander Krol, A.R.,J.N.M. Mol and R. Stuitje. 1988. *Gene,* 72: 45-50.

3

Pod-wall Metabolism in Relation to Dry Matter and Protein Accumulation in Developing Seeds of Chick-pea (*Cicer arietinum* L.)

V.K. Gupta and Randhir Singh

Department of Chemistry and Biochemistry,
Haryana Agricultural University, Hisar-125 004, India

ABSTRACT

Activities of some key enzymes of the PCR cycle and C_4 metabolism, rates of $^{14}CO_2$ fixation in the light and dark, and initial products of photosynthetic $^{14}CO_2$ fixation when determined in pod-wall and seed coat (fruiting structures) of chick-pea, indicated that compared to activities of RuBP carboxylase and other PCR cycle enzymes, the activities of PEP carboxylase and other enzymes of C_4 metabolism were generally much higher in fruiting structures than in the leaf. Pod-wall and seed coat fixed $^{14}CO_2$ in light and dark at much higher rates than the leaf. Short-term assimilation of $^{14}CO_2$ by illuminated fruiting structures produced malate as the major labelled product with less labelling in 3-PGA, whereas the lead showed major incorporation into 3-PGA.

PEP carboxylase purified to near homogeneity from the pod-wall, when analysed for physical and kinetic characteristics, indicated that this enzyme plays an important role in refixation of respired CO_2, which is subsequently released within the cell by the action of $NADP^+$-malic enzyme and could either be reduced via the PCR cycle or used in the generation of carbon skeletons (2-oxoglutarate) for the biosynthesis of amino acids. In the latter case, the oxaloacetate formed in the carboxylation step condenses with acetyl CoA to give ultimately isocitrate via citrate. Isocitrate dehydrogenase oxidatively decarboxylates isocitrate to 2-oxoglutarate, and the CO_2 released is recycled to form oxaloacetate, which could then be transaminated to form aspartate or follow the above sequence of reactions. In this way, recaptured, CO_2 by PEP carboxylase is effectively linked to the deposition of protein reserves in legume seeds. Various enzymes of this sequence, such as PEP carboxylase, $NADP^+$-malic enzyme, $NADP^+$-isocitrate dehydrogenase, pyruvate kinase and $NADP^+$-malate dehydrogenase, etc., have been purified and well characterised from the pod-wall of chick-pea. Based on these studies, we have proposed a metabolic pathway operating in the pod-wall of chick-pea which replenishes the intermediates of the TCA cycle and thus provides carbon skeletons for the biosynthesis of amino acids in chick-pea seeds.

Abbreviations: 3-PGA, 3-phosphoglyceric acid; PEP, Phosphoenol pyruvate; PCR, photosynthetic carbon reduction; M_r, molecular mass; KD, kilodalton; TCA, tricarboxylic acid; OAA, oxaloacetate; G-6-P, glucose-6-phosphate, F-1-6-bis P, fructose-1-6-bisphosphate.

INTRODUCTION

Developing pods of legumes are known to be photosynthetically active and are generally considered to perform a major role in carrying out the refixation of CO_2 released during either respiration or photorespiration [28]. However, the structural and carboxylation characteristics of pods vary widely depending on the legume species. In *Pisum arvense* L., the pod-wall supplies 66% of the carbon required by the developing seed during the period of maximum growth [8]. Compared to the leaf, more translocation of photosynthates has been observed to occur from the pod-wall during later-stages of pod development in pea [14]. Similarly, pod-wall of chick-pea has also been reported to contribute significant amounts of photosynthates to the developing fruit, and leaf photosynthesis could be complemented by pod-wall photosynthesis during pod development [15, 30]. However, compared to the leaf, detailed studies on the photosynthetic characteristics of the pod-wall and seed coat of legume crops have not yet been carried out during the entire period of seed development [4, 16, 19, 21].

Recent experiments, conducted in our laboratory in the presence of labelled $^{14}CO_2$, revealed that the first products of CO_2 fixation in the pod-wall of chick-pea, which utilises PEP carboxylase for the reaction, are oxaloacetate and malate [26], rather than 3-PGA, as is the case in the C_3 pathway. Based on these observations, it was postulated that the pod-wall surrounding the seed has the capacity for C_4 or CAM-like metabolism, thus helping to recapture respired or photo-respired CO_2. This recaptured CO_2 is subsequently released within the cell by the action of $NADP^+$- malic enzyme and could either be reduced via the PCR cycle or used in the generation of carbon skeletons (2-oxoglutarate) for amino acids [11]. In the latter case, oxaloacetate formed in the carboxylation step condenses with acetyl CoA to give ultimately isocitrate via citrate. Isocitrate is then oxidatively decarboxylated by $NADP^+$-isocitrate dehydrogenase to 2-oxoglutarate. The released CO_2 is recycled to form oxaloacetate. In recent studies in this laboratory, detailed investigations were conducted on the carbon fixation characteristics of the pod-wall of chick-pea [24] and the various enzymes involved in the generation of carbon skeletons characterized [7, 11, 13, 25], which are needed for the biosynthesis of amino acids in the developing seeds. Based on these studies, a pathway operating in the pod-wall of chick-pea replenishing the intermediates of TCA cycle and thus providing carbon skeletons for the biosynthesis of amino acids in seeds, were postulated.

MATERIALS AND METHODS

Chemicals

All the biochemicals and enzymes were purchased from Sigma (St. Louis, MO, USA). NaH $^{14}CO_3$ (sp activity 1.85 TB_q mol^{-1}) was purchased from Bhaba Atomic Research Centre (India). All other chemicals used were of analytical grade.

Plant Material

A chick-pea crop (cv H75-35) was raised in the fields of the Pulse Section of Haryana Agricultural University, Hisar according to the recommended agronomic practices. Fully opened flowers were tagged on the day of anthesis.

Twenty pods, sampled at random at 3-day intervals starting from day 3 after anthesis up to full maturity, were brought to the laboratory and their fresh and dry weights recorded after separating the pods into pod-walls and seeds. Carbon dioxide exchange studies were carried out at 3-day intervals by following the procedure described elsewhere [24]. During the period of rapid seed growth (20 days after anthesis), $^{14}CO_2$ feeding was done both from outside and inside the pod-wall as described earlier [24]. After 20 sec incubation in the presence of labelled CO_2, the pods were sampled immediately after feeding and at 60, 120, 300 sec and 24 and 48 h after the start of feeding. The pods were separated into pod-walls and seeds and killed in boiling 80% ethanol. ^{14}C distribution in the ethanol soluble fraction was determined by liquid scintillation counter as done earlier [26]. Internal CO_2 concentration of pods was measured with an infrared gas analyser operating in a differential mode. These measurements were made in full sunlight (> 1200 μe m^{-2}).

Products of $^{14}CO_2$ assimilation: Immediately after separation, the tissues were incubated with $^{14}CO_2$(500 μ 1 L^{-1}) in a perspex chamber for 20, 40, 60, 120 and 300 sec, using the technique described previously [26]. After incubation, the tissues were immediately killed and extracted in boiling 80% ethanol. The ethanol extract was evaporated to dryness. Chlorophyll was extracted from the solids by washing twice with chloroform. After evaporating excess chloroform, the solids were taken up in water. A suitable aliquot of the water extract was then subjected to paper chromatography using n-butanol-propionic acid-water (10:5:7, v/v) as the solvent [2]. Radioactive products were identified by co-chromatography and autoradiography. Areas containing ^{14}C were punched from the paper, extracted in 80% ethanol and radioactivity determined by liquid scintillation counting (Beckman LS 100C) with an efficiency of 80%.

Enzyme extraction: Five hundred mg of each tissue were used and enzyme extracts prepared as described previously [27].

Enzyme assays: Enzyme activities were determined spectrophotometrically at 340 nm by following the oxidation of NAD (P) H or reduction of NADp (P)⁺. All assays were carried out at 30°C. Preliminary assays were done for all the enzymes to determine optimum conditions where linear reaction rates with respect to time and enzyme concentrations were obtained. The activities of various key enzymes of the PCR cycle, viz., RuBP carboxylase (EC 4.1.1.39), NADP⁺-glyceraldehyde 3-phosphate dehydrogenase (EC 1.2.1.13), and ribulose 5-phosphate kinase (EC 2.7.1.19) and C₄ metabolism, viz., PEP carboxylase (EC 4.1.1.31), NAD⁺–malate dehydrogenase (EC 1.1.1.37), NADP⁺–malate dehydrogenase (EC 1.1.1.82), NAD⁺–malic enzyme (EC 1.1.1.39), NADP⁺-malic enzyme (EC 1.1.1.40), glutamate oxaloacetate transaminase (EC 2.6.1.1) and glutamate pyruvate transaminase (EC 2.6.1.2), were determined by following the standard assay methods as reported earlier [26].

Enzyme purification: PEP carboxylase [25], NADP⁺-malic enzyme [7], NADP⁺-isocitrate dehydrogenase [11], pyruvate kinase [12] and NADP⁺-malate dehydrogenase [13] from immature chick-pea pod-wall were purified partially by following the procedures contained in the foregoing references.

Estimation of metabolites: Unless otherwise stated, all operations were carried out at 0°C. Metabolites were extracted by following the method of Adams and Rinne [1] and estimated quantitatively by use of coupled assays by following the reduction of NAD(P) or oxidation of NAD (P) H at 340 nm in a Calbiochem Enzymeter. Pyruvate and PEP were assayed as per the procedure of Czok and Lamprecht [5]. Oxaloacetate, malate, citrate, 2-oxoglutarate and alanine were measured by following the methods of Rej [22], Mollering [20], Dagley [6], Burlina [3] and Grassl [10] respectively.

Estimation of protein and chlorophyll. Protein in the enzyme extracts was measured by following the method of Lowry and co-workers [18] after precipitation with trichloroacetic acid. Chlorophyll was estimated according to Strain and co-authors [31].

RESULTS

Pod Development

Pod development in chick-pea can be divided into two phases; the first comprises pod-wall development and continues up to day 15 after anthesis, and the second phase is that of seed development starting from day 15 after anthesis to day 42 after anthesis. The dry weight of the pod-wall increased up to day 15 after anthesis and seed growth started after this period was maximum during the period 30 to 39 days after anthesis (Fig. 1).

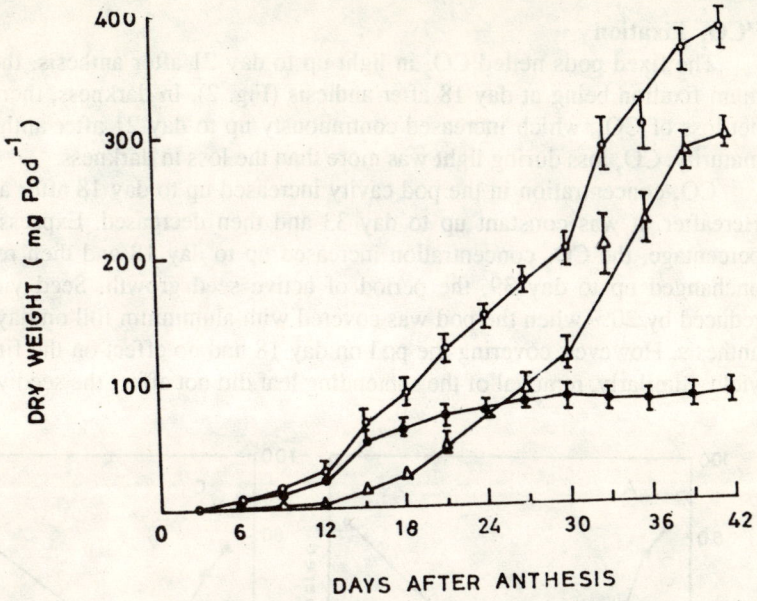

Fig. 1: Dry weight of pod (O—O), seed (△—△) and
pod-wall (●—●) at different stages after anthesis.

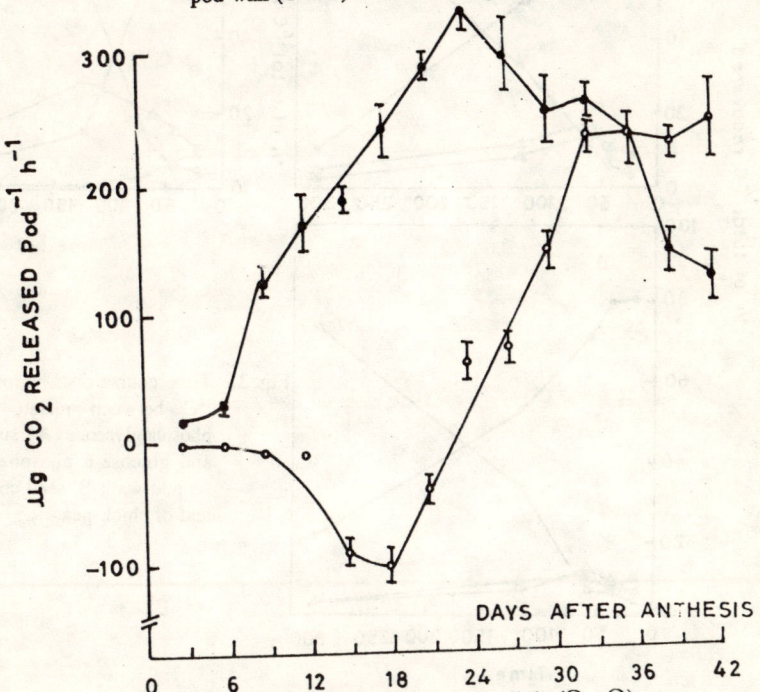

Fig. 2: Net CO_2 exchange by pod during light (O—O)
and dark (●—●) at different stages after anthesis.

$^{14}CO_2$ Fixation

The fixed pods netted CO_2 in light up to day 21 after anthesis, the maximum fixation being at day 18 after anthesis (Fig. 2). In darkness, there was a net loss of CO_2, which increased continuously up to day 21 after anthesis. At maturity, CO_2 loss during light was more than the loss in darkness.

CO_2 concentration in the pod cavity increased up to day 18 after anthesis. Hereafter, it was constant up to day 33 and then decreased. Expressed as a percentage, the CO_2 concentration increased up to day 18 and then remained unchanged up to day 39, the period of active seed growth. Seed yield was reduced by 20% when the pod was covered with aluminium foil on day 3 after anthesis. However, covering the pod on day 18 had no effect on the final seed yield. Similarly, removal of the subtending leaf did not affect the seed yield.

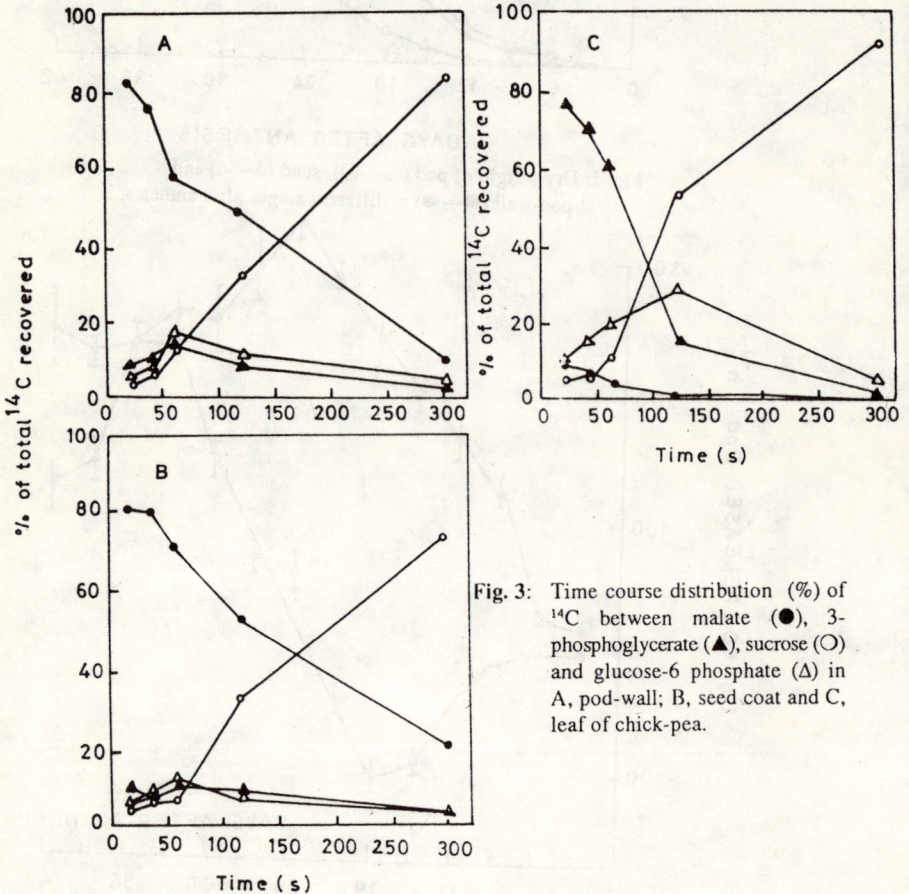

Fig. 3: Time course distribution (%) of ^{14}C between malate (●), 3-phosphoglycerate (▲), sucrose (○) and glucose-6 phosphate (Δ) in A, pod-wall; B, seed coat and C, leaf of chick-pea.

No labelling was observed in the seeds up to 5 min, when $^{14}CO_2$ was fed from outside. Only at 48 h, did the seeds contain about 50% of the label. On the contrary, seeds fixed only a small amount of CO_2 (10%) when the pod was fed through the internal cavity. In the latter case, fixation occurred mainly in the pod-wall.

The distribution of radioactivity between malate, 3-PGA, sucrose and glucose-6 P obtained as products of $^{14}CO_2$ assimilation, was determined after 20,40,60,120 and 200 sec of incubation. Of the $^{14}CO_2$ fixed after 20 sec of photosynthesis in the pod-wall, seed coat and leaf, about 82, 80 and 8% was in malate, whereas 9, 10 and 77% respectively was in 3-PGA (Fig. 3). The remaining label was in sucrose and glucose-6-P. The ratio of malatae to 3-PGA was 9.1, 8.0 and 0.1 in the pod-wall, seed coat and leaf respectively. After 120 sec, while sucrose was the major labelled product in the leaf, the level of labelled malate still remained higher than that of sucrose in spite of an increase in the level of labelled sucrose and a decrease in labelled malate. At 300 sec, sucrose accounted for 70 to 90% of the total ^{14}C recovered.

Enzymes of PCR Cycle and C_4 Metabolism

The activity of RuBP carboxylase, a key enzyme of the PCR cycle, was higher in the leaf than in the pod-wall and seed coat at all stages of seed development (Table 1). The enzyme activity was highest in the youngest tissues

Table 1: Enzyme activities in pod-wall of chic-kpea during seed development

Enzyme	Enzyme activity (nmol min^{-1} mg^{-1} protein)			
	10	20	30	40
RuBP-carboxylase	73 ± 0.8	93 ± 13	55 ± 0.4	2.0 ± 0.1
NADP$^+$-glyceraldehyde 3-phosphate dehydrogenase	257 ± 13.9	299 ± 24.7	293 ± 4.6	14.0 ± 0.3
Ribulose-5-phosphate kinase	915 ± 13.4	1638 ± 24.9	1025 ± 14.9	750 ± 0.6
PEP carboxylase	74 ± 7.1	99 ±8.0	159 ± 2.7	27 ± 0.2
NADP$^+$-malate dehydrogenase	30 ± 0.2	40 ± 0.6	122 ± 1.4	19 ±0.2
NADP$^+$-malic enzyme	37 ± 0.4	46±0.4	94 ± 1.1	22 ± 0.5
Glutamate oxaloacetate transaminase	234 ± 3.7	386 ± 5.2	433 ± 5.9	103 ±1.1
Glutamate pyruvate transaminase	139 ± 1.8	153 ± 1.9	224 ± 3.1	62 ± 1.0
Pyruvate kinase	107 ± 1.2	231 ± 3.3	153 ± 1.9	68 ± 0.7

and declined as maturity advanced. However, in the pod-wall, the enzyme activity declined only after 20 days of flowering. The other PCR cycle enzymes examined, viz., NADP-glyceraldehyde-3-phosphate dehydrogenase, NAD+-glyceraldehyde-3-phosphate dehydrogenase and ribulose-5-phosphate kinase, showed patterns qualitatively similar to that of RuBP carboxylase. Ribulose-5-phosphate kinase activity, however, was very high compared to the other enzymes of the PCR cycle at all stages.

Contrary to RuBP carboxylase, PEP carboxylase was more active in the pod-wall and seed coat than in the leaf at all stages of seed development [29]. The enzyme activity in the leaf and seed coat decreased as maturity advanced, whereas in the pod-wall it first increased up to 30 days after flowering and decreased thereafter. The other enzymes of C_4 metabolism investigated, NADP+-malate dehydrogenase, NAD+-malic enzyme, glutamate oxaloacetate transaminase, and glutamate pyruvate transaminase, followed a pattern similar to that for PEP carboxylase.

Enzyme Properties

PEP carboxylase (EC 4.1.1.31) was purified to homogeneity with about 29% recovery from immature pods of chick-pea using ammonium sulphate fractionation, DEAE-cellulose chromatography, and gel filtration through Sephadex G-200 [25]. The purified enzyme with mol wt of about 200,000 daltons was a tetramer of four identical subunits and exhibited maximum activity at pH 8.1 (Table 2). Mg^{2+} ions were specifically required for the enzyme activity. The enzyme showed typical hyperbolic kinetics with PEP with a Km of 0.74 mM, whereas a sigmoidal response was observed with increasing concentrations of HCO_3^- with $S_{0.5}$ value as 7.6 mM. The enzyme was activated by inorganic phosphate and phosphate esters such as glucose-6-P, α-glycerophosphate, 3-PGA, and fructose-1,6-P_2, and inhibited by nucleotide triphosphates, organic acids, and divalent cations Ca^{2+} and Mn^{2+}. Oxaloacetate and malate inhibited the enzyme non-competitively. Glucose-6-P reversed the inhibitory effects of oxaloacetate and malate.

NADP+-malic enzyme (EC 1.1.1.40) was purified 51-fold by ammonium sulphate fractionation, DEAE-cellulose chromatography and gel filtration through Sepharose 4B [7]. The purified enzyme required a divalent cation, either Mn^{2+} or Mg^{2+}, for its activity (Table 2). Km values at pH 7.8 for malate, NADP+ and Mn^{2+} were 4.0, 0.031 and 0.71 mM respectively. Mn^{2+}-dependent activity was inhibited by heavy metal ions such as CD^{2+}, Zn^{2+}, Hg^{2+}, and to a lesser extent by Pb^{2+} and Al^{2+} and Al^{3+}. Among the organic acids examined, sodium salts of oxalate and oxaloacetate were inhibitory. Kinetics of the reaction mechanism showed sequential binding of malate and NADP+ to the enzyme. Products of the reaction, viz., pyruvate, bicarbonate and NADPH, inhibited the enzyme activity. At limiting concentrations of NADP+, pyruvate and bicarbonate induced a positive co-operative effect by malate. Based on these

Table 2: Kinetic characteristics of important enzymes involved in carbon metabolism in pod-wall of chick-pea

Enzyme	Approx. mol wt (kD)	Optimum	Km/S*0.5		Metal ion requirement	Effectors		Mechanism of action
			Substrate	mM		+ve	-ve	
PEP carboxylase	200	8.1	PEP HCO$_3$'	0.740 7.600*	Mg^{2+}	Pi 3-PGA G-6-P F-1,6-bis P a -Glycero-phosphate	OAA malate ATP CTP UTP	ND
Pyruvate kinase	200	6.8	PEP ADP	0.100 0.174	Mg^{2+} K$^+$	AMP GMP CMP F-1,6-bis P	ATP UTP Citrate glutamate oxalate	Sequential
NADP-malate dehydrogenase	135	6.0-7.0	OAA NADPH malate NADP+	0.062 0.011 2.000 0.290	NIL	NIL	Citrate isocitrate oxalate 2-oxoglutarate ATP malate NADP*	Sequential (ordered)
NADP-malic enzyme	ND	7.8	malate NADP	4.00 0.031	Mg^{2+} Mn^{2+}	NIL	Oxalate OAA	Sequential (ordered)
NADP-isocitrate dehydrogenase	126	8.0-8.6	NADP isocitrate	0.015 0.110	Mg^{2+}/ Mn^{2+}	NIL	NADH NADPH 2-oxoglutarate	Sequential (random)

Note: ND means not determined.

* The enzyme showed sigmoidal response to increasing concentrations of HCO$_3$.

properties, it was proposed that the activity of this enzyme is controlled by intracellular concentrations of substrates and products.

NADP+-malate dehydrogenase (EC 1.1.1.82) partially purified as above had an M_r of about 135 kD and exhibited optimum pH between 6.0-7.0 [13]. The enzyme showed normal hyperbolic response for both the pairs of substrates in forward and reverse reactions with Km values of 0.062 and 0.11 mM for oxaloacetate and NADPH, and 2.0 and 0.29 mM for malate and NADP+ respectively (Table 2). The enzyme activity in the forward direction was inhibited by citrate, isocitrate, oxalate, 2-oxoglutarate, ATP, malate and NADP. The Ki values for isocitrate and oxalate was 10 mM and 4.5 mM respectively. Initial velocity studies indicated sequential binding of substrates to the enzyme. Product inhibition studies suggested that substrates bind and products leave the enzyme in an ordered manner.

NADP+-isocitrate dehydrogenase (EC 1.1.1.42) purified to 192-fold had a mol wt of about 126 kD [11]. The enzyme exhibited a broad pH optima from 8.0 to 8.6. It was quite stable at 4°C and had an absolute requirement for a divalent cation, either Mg^{2+} or Mn^{2+} for its activity (Table 2). Typical hyperbolic kinetics was obtained with increasing concentrations of NADP+, DL-isocitrate, Mn^{2+} and Mg^{2+}. Their Km values were 15, 110, 15 and 192 µM respectively. The enzyme activity was inhibited by sulphydryl reagents. Various amino acids, amides, organic acids, nucleotides, each at a concentration of 5 mM, had no effect on the activity of the enzyme. The activity was not influenced by adenylate energy charge but decreased linearly with an increasing ratio of NADPH to NADP+. Initial velocity studies indicated the kinetic mechanism to be sequential. NADPH inhibited the forward reaction competitively with respect to NADP+ at fixed saturating concentrations of isocitrate, whereas 2-oxoglutarate inhibited the enzyme non-competitively at saturating concentrations of both NADP+ and isocitrate. These results suggest that the activity of NADP+-isocitrate dehydrogenase in situ is probably controlled by the intracellular NADPH to NADP+ ratio as well as by the concentration of various substrates and products.

Pyruvate kinase (EC 2.7.1.40) partially purified as above had an M_r of 200 kD and exhibited pH optimum near 6.8 [12]. Typical Michaelis-Menten kinetics was obtained for both the substrates with Km values of 0.10 and 0.17 mM for PEP and and ADP respectively (Table 2). The enzyme could also utilize UDP or GDP but with lower V_{max} and lesser affinities. It had an absolute requirement for a divalent cation, preferably Mg^{2+}, and a monovalent cation, K+, for optimal activity. The enzyme was inhibited by ATP, citrate, UTP, glutamate, and oxalate and activated by AMP, GMP, CMP and fructose-1, 6-P_2. Initial velocity studies indicated sequential binding of the substrates to the enzyme. Based on these characteristics, it was proposed that pyruvate kinase, besides participating in respiratory control might also help in supplying additional carbon skeletons for ammonia assimilation.

Table 3: Concentration of various metabolites involved in the metabolism of PEP in pod-wall of chick-pea at different stages of seed development

DAF	Metabolite						
	OAA	Pyruvate	PEP	2-oxoglutarate	Alanine	Malate	Citrate
	μmol/g dry wt					μmol/g dry wt	
10	417 ± 12.4	436 ± 8.6	410 ± 16.0	1310 ± 36.4	4670 ± 48.0	251.73 ± 1.00	18.20 ± 0.44
17	273 ± 9.6	356 ± 10.2	227 ± 10.4	431 ± 16.2	704 ± 26.6	67.22 ± 0.68	7.55 ± 0.30
24	185 ± 6.0	282 ± 12.0	189 ± 9.0	403 ± 14.8	1169 ± 64.1	75.64 ± 1.36	1052 ± 0.38
31	377 ± 16.4	582 ± 14.6	246 ± 12.2	631 ± 23.4	2224 ± 76.2	83.51 ± 1.44	10.75 ± 0.48
38	286 ± 8.2	238 ± 17.4	300 ± 14.6	416 ± 16.0	441 ± 28.4	100.90 ± 1.08	40.3 ± 1.60

Levels of Metabolites

The pod-wall contains relatively small amounts of PEP, pyruvate, alanine, oxaloacetate and 2-oxoglutarate, with substantial amounts of citrate and malate (Table 3). The levels of all these metabolites were highest at day 10 after anthesis. The levels of malate and PEP increased from day 17 after anthesis onwards while those of oxaloacetate, pyruvate, citrate, 2-oxoglutarate and alanine increased up to day 31 after anthesis and decreased thereafter.

DISCUSSION

Developing pods of legumes are known to be photo-synthetically active [28]. However, the extent of their contribution towards seed growth is still not clear. Though a number of studies conducted earlier have indicated that legume pods fix very little net CO_2 in light [4, 21], in the present investigation a net fixation of CO_2 occurred in light, which increased up to day 18 after anthesis (Fig. 2), indicating that the pod-wall of chick-pea fixes atmospheric CO_2 photosynthetically during the early phase of pod development. This was further confirmed by labelling experiments wherein pod-wall fixed $^{14}CO_2$ when fed externally. In the present case, pod-wall through mobilisation and net CO_2 fixation contributed about 20% of the photosynthates towards seed dry weight. This is contrary to the results reported by Singh and Pandey [30]; they concluded that pod-wall photosynthesis was not a significant source of assimilate for seed development. In pea also, pod photosynthesis was shown to improve the economy of carbon usage by about 16 to 20% [9].

The chick-pea pod-wall was far more efficient (90%) in $^{14}CO_2$ fixation than the seeds (10%) when fed through the internal cavity. Since no stomata was observed on the inner layer of the pod-wall, the source of CO_2 for this could only be the respired CO_2. In this way, the pod-wall acts as an impermeable barrier to internal CO_2, accumulating a very high amount in the pod-cavity, especially during the rapid phase of seed growth. Interestingly, seeds did not develop in this crop when the pod-wall was made permeable by puncturing and allowing CO_2 to escape.

Two experimental approaches, namely kinetic studies with $^{14}CO_2$ and enzyme profiles, were implemented to determine further the pathway of CO_2 assimilation in fruiting structures of chick-pea. Interestingly, in the pod-wall and seed coat, the initial product of $^{14}CO_2$ assimilation after 20 sec of photosynthesis was malate. About 80% of the total radioactivity appeared in malate compared to about 10% in 3-PGA (Fig.3). However, the leaf showed 77% of radioactivity in 3-PGA and only 8% in malate. The rapid labelling and relatively high proportion of ^{14}C in malate seems to reflect the synthesis of OAA catalysed by PEP carboxylase. These results were supported by enzymic studies. The activities of RuBP carboxylase and other PCR cycle enzymes were high in the leaf compared to the pod-wall and seed coat (Table 1), which had

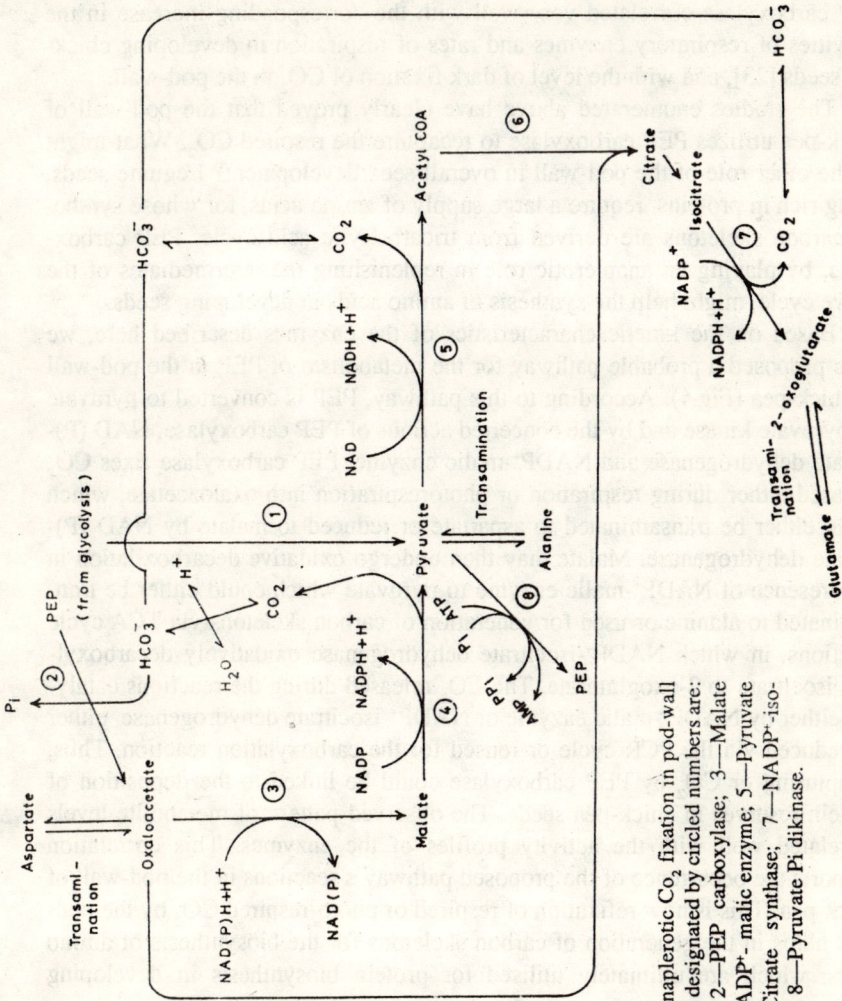

Fig. 4: Probable pathway of anaplerotic CO_2 fixation in pod-wall of chick-pea. Enzymes designated by circled numbers are: 1—Pyruvate kinase; 2—PEP carboxylase; 3—Malate dehydrogenase; 4—NADP⁺- malic enzyme; 5—Pyruvate dehydrogenase; 6—Citrate synthase; 7—NADP⁺-isocitrate dehydrogenase; 8—Pyruvate Pi dikinase.

higher levels of PEP carboxylase and other enzymes of 4 metabolism at all stages of seed development. The observed increase in the level of pod-wall PEP carboxylase correlated very well with the corresponding increase in the activities of respiratory enzymes and rates of respiration in developing chick-pea seeds [23], and with the level of dark fixation of CO_2 in the pod-wall.

The studies enumerated above have clearly proved that the pod-wall of chick-pea utilizes PEP carboxylase to recapture the respired CO_2. What might be the other role of the pod-wall in overall seed development? Legume seeds, being rich in proteins, require a large supply of amino acids, for whose synthesis carbon skeletons are derived from tricarboxylic acid cycle. PEP carboxylase, by playing an anaplerotic role in replenishing the intermediates of the above cycle, might help the synthesis of amino acids in developing seeds.

Based on the kinetic characteristics of the enzymes described here, we have proposed a probable pathway for the metabolism of PEP in the pod-wall of chick-pea (Fig.4). According to this pathway, PEP is converted to pyruvate by pyruvate kinase and by the concerted actions of PEP carboxylase, NAD (P)-malate dehydrogenase and $NADP^+$-malic enzyme. PEP carboxylase fixes CO_2 released either during respiration or photorespiration into oxaloacetate, which could either be transaminated to aspartate or reduced to malate by NAD (P)-malate dehydrogenase. Malate may then undergo oxidative decarboxilation in the presence of $NADP^+$-malic enzyme to pyruvate which could either be transaminated to alanine or used for generation of carbon skeletons via TCA cycle reactions, in which $NADP^+$-isocitrate dehydrogenase oxidatively decarboxylates isocitrate to 2-oxoglutarate. The CO_2 released during the reactions catalysed either by $NADP^+$-malic enzyme or $NADP^+$-isocitrate dehydrogenase either be reduced via the PCR cycle or reused for the carboxylation reaction. Thus, recapturing or CO_2 by PEP carboxylase could be linked to the deposition of protein reserves in chick-pea seeds. The observed pattern of metabolite levels correlated well with the activity profiles of the enzymes. This correlation supports the occurrence of the proposed pathway's reactions in the pod-wall of chick-pea. This is how refixation of respired or photo-respired CO_2 by the pod-wall helps in the generation of carbon skeletons for the biosynthesis of amino acids, which are ultimately utilised for protein biosynthesis in developing seeds.

Acknowledgements

Financed in part by a grant from the United States Department of Agriculture under the Co-operative Agricultural Research Grant Program (PL-480). The senior author is thankful to the Council of Scientific and Industrial Research, New Delhi for a junior research fellowship.

LITERATURE CITED

1. Adams, C.A. and R.W. Rinne. 1981. Interaction of phosphenolypyruvate carboxylase and pyruvic kinase in developing soybean seeds. *Plant and Cell Physiol;* **22**: 1011-1021.
2. Benson, A.A, J.A. Bassham M. Calvin, T.C, Goodale, V.A. Mass and W. Stepka, 1950. The path of carbon in photosynthesis. V. Paper chromatography and radioautography of the products. *J. Am. Chem. Soc.,* **72**: 1710-1718.
3. Burlina, A. 1985. 2-oxoglutarate. In: *Methods of Enzymatic Analysis* edited by H.U. Bergmeyer. VCH, Verlagsgesellschaft, Weinheim, vol. VII, pp. 20-24.
4. Crookston, R.K., J. O'Toole and J.L. Ozbun, 1974. Characterization of the bean pod as a photosynthetic organ. *Crop. Sci.,* **14**: 708-712.
5. Czok, R. and W. Lamprecht, 1974. Pyruvate, phosphenol pyruvate and D-glycerate-2 phosphate. In: *Methods of Enzymatic Analysis,* edited by H.U.Bergmeyer. Verlag Chemie, Weinheim, vol. III, PP. 1446-1451.
6. Dagley, S. 1974. Citrate, UV spectrophotometric determination. In: *Methods of Enzymatic Analysis*, edited by H.U.
7. Das, S., D.R. Sood, S.K. Sawhney and R. Singh 1986. Properties of NADP⁺-malic enzyme from pod-walls of chick-pea. *Physiol. Plant* 68: 308-314.
8. Flinn, A.M. and J.S. Pate, 1970. A quantitative study of carbon transfer from pod and subtending leaf to the ripening seeds of the field pea. *J. Exp. Bot;* 21: 71-82.
9. Flinn, A.M, C.A. Atkins and J.S. Pate 1977. Significance of photosynthetic and respiratory exchanges in the carbon economy of the developing fruit. *Plant Physiol.* 60: 412-418.
10. Grassl, M. 1974. L-alanine: determination with GTP and LDH. In: *Methods of Enzymatic Analysis*, edited by H.U. Bergmeyer Verlag Chemie, Weinheim, vol. IV, pp. 1682-1685.
11. Gupta, V.K. and R. Singh.1988. Partial purification and characterization of NADP⁺- isocitrate dehydrogenase from immature pod-walls of chick-pea. *Plant Physiol.,* **87**: 741-744.
12. Gupta, V.K. and R. Singh, 1989a. Properties of NADP-mantle dehydrogenase from immature pod-wall of chick-pea. *Physiol. Plant* (in press).
13. Gupta, V.K. and Singh, 1989b. Properties of pyruvate kinase from immature pod-wall of chick-pea. *Plant Physiol.,* (in press).
14. Khanna-Chopra, R. and S.K. Singh 1976. Importance of fruit wall in seed yield of pea (*Pisum sativum* L.) and mustard.(*Brassica campestris* L.). *Indian J. Exp. Biol.* **14**: 159-162.
15. Khanna-Chopra, R. and S.K. Sinha, 1982. Photosynthetic rate and photosynthetic carboxylation enzymes during growth and development in *Cicer arietinum* L. cultivars. *Photosynthetica* **16**: 509-513.
16. Koundal, K.R. and S.K. Sinha, 1987. Malic acid exudation and photosynthetic characteristics in *Cicer arietinum. Phytochemistry,* **20**: 1251-1252.
17. Latzko , E. and G.J. Kelley, 1983. The many-faceted function of phosphenolpyruvate carboxylase in C₃ plants. *Physiol. Veg.* **21**:805-815.
18. Lowry, O.H. N.J. Rosebrough, A.L. Farr and RL Randall, 1951. Protein measurement with the Folin phenol reagent. *J. Biol. Chem.* **193**: 265-275.
19. Luthra, Y.P., I.S. Sheoran and R. Singh, 1983. Photosynthetic rates and enzyme activities of leaves, developing seeds and pod-wall of pigeonpea (*Cajanus cajan* L.). *Photosynthetica,* **17**: 210-215.
20. Mollering, H. 1974. L (-) malate-determination with MDH and GOT. In; *Methods of Enzymatic Analysis* edited by H.U. Bergmeyer Verlag Chemie, Weinheim Vol. III, pp. 1589-1593.
21. Quebedeaux, and B.R. Choilet 1975. Growth and development of soybean pods: CO₂ exchange and enzyme studies. *Plant Physiol.,* **55**: 745-748.

22. Rej, R. 1985. Oxaloacetate: UV-method. In: *Methods of Enzymatic Analysis* edited by H.U. Bergmeyer. V.C.H., Verlagsgesellschaft, Weinheim, Vol. VII, pp. 59-67.

23. Sangwan, R.S., S. Popli, and R. Singh, 1983. Carbohydrate oxidising enzymes and rate of oxygen consumption in developing chick-pea seeds. *Indian J. Exp Biol* 21: 37-39.

24. Sheoran, I.S. H.R. Singhal and R. Singh, 1987. Photosynthetic characteristics of chick-pea pod-wall during seed development. *Indian J. Exp. Biol.*, 25: 843-847.

25. Singal, H.R. and R. Singh. 1986. Purification and properties of phosphoenolpyruvate carboxylase from immature pods of chickpea (*Cicer arietinum* L.). *Plant Physiol.*, 80: 369-373.

26. Singal, H.R. I.S. Sheoran and R.Singh, 1986a. Products of photosynthetic $^{14}CO_2$ fixation and related enzyme activities in fruiting structures of chickpea. *Physiol. Plant.*, 66: 457-462.

27. Singal, H.R., I.S. Sheoran and R. Singh. 1986.b. *In vitro* enzyme activities and products of $^{14}CO_2$ assimilation in flag leaf and ear parts of wheat (*Triticum aestivum* L.). *Photosynthesis* Research 8: 113-122.

28. Singh, R., 1987. Photosynthetic carbon fixation characteristics of fruiting structures of wheat, chick-pea and *Brassica* sp. *Proc. Indian Natn. Sci. Acad.* 53: 541-544.

29. Singh, R., 1989. Carbon dioxide fixation by PEP carboxylase in pod-walls of chick-pea. (*Cicer arietinum*) In: Proc. Int. workshop on Applications of Molecular Biology and Bioenergetics of Photosynthesis. edited by G.S. Singhal, J. Barber, R.A. Dilley, Govindjee, R. Hasekorn and P. Mohanty, Narosa Publishing House, pp. 315-329.

30. Singh, B.K. and R.K. Pandey, 1980. Production and distribution of assimilate in chick-pea (*Cicer arietinum*). *Aust. J. Plant Physiol.*, 7:727-735.

31. Strain, H.H., B.T. Cope and A.R. Svec. 1971. Analytical procedures for the isolation, identification, estimation and investigation of the chlorophylls. *Methods Enzymol.*, 23: 452-476.

4

Micropropagation of Woody Taxa and Plant Productivity

Shrish C. Gupta and Veena Agrawal

Department of Botany, University of Delhi, Delhi-110007, India

ABSTRACT

Regeneration of nine woody taxa, i.e. *Acacia auriculiformis, A. nilotica, Albizia amara, A. lucida, A. procera, A. richardiana, Leucaena leucocephala, Sesbania macrocarpa* and *Populus × euramericana* has been accomplished. The explants were excised from *in vitro* raised seedlings in the first five species, field-grown as well as juvenile tissues in the next three and 20 to 25-year-old trees in the last taxon (*Populus*). The axillary buds differentiated multiple shoots in *Acacia auriculiformis* [32] and *A. nilotica* (Gupta, unpublished) on B_5 + BA + CW medium. Shoots of the former taxon, after subsequent transfer to B_5+ NAA or IAA medium, organised roots [32]. Plantlets developed in hypocotyl explants of all the four species of *Albizia* through organogenesis. Plants also differentiated *via* somatic embryogenesis in *A. richardiana* [48-51]. In *Leucaena leucocephala*, plantlets were obtained from cotyledons and axillary buds on B_5 + BA medium and from shoot tips of 5-year-old field-grown trees on MS + BA + NAA medium [22]. Both the nodal and internodal segments of *Populus* differentiated multiple shoots even on MS basal medium. Its excised shoots rooted well on MS + NAA + AC medium. In this species, seasonal variation in morphogenic response of stem explants is very conspicuous. Maximum response (up to 95%) was observed from March to May, which declined gradually and dropped to zero from December to February. In *Sesbania macrocarpa*, multiple shoots developed from cotyledon, hypocotyl and nodal explants on B_5+ BA medium. Such excised shoots organised roots on B_5 + IBA. The plantlets were transferred to the field where they flowered and set viable seeds [24, 25]. In addition, leaflet explants of *Sesbania* also organised 15-20 shoots per explant on B_5 + BA + NAA + PVP medium.

INTRODUCTION

Continuous deforestation has created an unprecedented scarcity of food, fodder, fuel and raw material for forest-based industries, such as paper, timber, match and pharmaceuticals, etc. throughout the world. To cope with these challenging problems, it is imperative to evolve methods by which fast-growing trees, capable of producing increased quantities of biomass, could be propagated. Plant, cell and tissue culture technology has undoubtedly emerged as a potential tool for rapid clonal propagation of rare and elite trees, developing mutant trees, increasing bio-mass energy and maintaining the desired gene bank [4, 8, 17]. For a successful large-scale propagation, however, three major problems must be overcome: (i) establishment of primary explants in culture, (ii) development of optimal culture media and environmental conditions to achieve a high rate of multiplication, and (iii) induction of roots as well as acclimatisation of plantlets during transfer to soil [35].

Though considerable attention has been paid to the micropropagation of several tree taxa [4, 10, 15], yet a number of woody species, especially the legumes, remain unexplored. Some reports on the regeneration of trees have appeared recently for *Dalbergia latifolia* from shoot tip explants [47], *Albizia lebbeck* from hypocotyl, leaf and stem segments [41], *Eucalyptus camaldulensis* from leaf explants [33], different explants excised from various species of juvenile as well as field-grown plants of *Populus* [1, 14, 28, 29, 38, 44], *Alnus cordata* from axillary buds [5], etc.

This paper embodies the various *in vitro* micropropagation techniques standardised by our research group to achieve successful regeneration in the following nine woody taxa: *Acacia auriculiformis* [32], *A. nilotica* (Gupta, unpublished), *Albizia amara, A. lucida, A. richardiana* [48, 51], *A. procera* (Goel and Gupta, unpublished), *Leucaena leucocephala* [22], *Populus* × *euramericana* and *Sesbania macrocarpa* [24, 25].

Besides micropropagation through organogenesis, the induction of somatic embryogenesis and the subsequent regeneration of plantlets in *Albizia* spp. [48-50] are also discussed.

MATERIALS AND METHODS

Plant Materials: Seeds of leguminous taxa were obtained from a seed nursery and surface-sterilised with freshly prepared chlorine water (3.5 ± 0.5 g.l^{-1}) for 20-50 min and washed subsequently 4 or 5 times with sterilised distilled water before inoculation. The seeds were germinated aseptically on Knop's modified medium [51] containing 0.03 M sucrose. The medium was gelled with 0.5-1% agar (Centron, India) and the pH was adjusted to 5.8 before autoclaving. The seeds were reared in culture tubes (Corning/Borosil) on 20 ± 2 ml nutrient medium per tube and incubated under continuous light ranging from 20 to 640 μW . cm^{-2} at 25 ± 2° C and 55 ± 10% relative humidity (RH). Explants were cut from *in vitro* raised seedlings of all the leguminous taxa as well as field-grown selected plants of *Albizia richardiana, Leucaena leucocephala* and *Sesbania macrocarpa.*

Small twigs (5-10 cm long) of *Populus × euramericana* were cut from 20 to 25-year-old trees growing in Delhi. After removing the leaves, the stem segments were treated with 5% (v/v) polysan solution for 10 min, washed thoroughly under running tap water and transferred to 70% ethanol for 10-15 min. As the chlorine water was ineffective, the material was surface-sterilised with 0.1% HgCl$_2$ for 10-15 min and finally washed 3 or 4 times with sterile distilled water. The youngest fourth to seventh internodes and third to eighth nodes were selected for inoculation.

Culture Media: Various basal media [21, 34, 54] were tried for each taxon and the one giving an optimal response was selected for further experiments. B$_5$ medium was chosen for culturing all the four species of *Albizia, Acacia auriculiformis, A. nilotica* and *Sesbania macrocarpa*. MS and B$_5$ media were found effective for *Leucaena leucocephala*. MS medium was better for *Populus*. The medium contained 0.09 M sucrose (Central Drug House, Bombay, India), unless otherwise mentioned, and gelled with 0.8-0.9% agar (Japan). The medium was supplemented with various auxins, such as IAA (indole-3-acetic acid), IBA (indole-3-butyric acid), NAA (α-naphthaleneacetic acid) and cytokinins, viz., BA (benzyladenine) and Kn (kinetin), in different concentrations individually or in combinations, as and when required. In addition, CW (coconut water) was also used for *Acacia auriculiformis* explants. PVP (polyvinyl pyrrolidone) was employed for controlling browning of tissues of *Leucaena leucocephala* and *Sesbania macrocarpa*. The addition of activated charcoal to the auxin-supplemented medium was necessary for the induction of rhizogenesis in *Populus*. The pH of the media was adjusted to 5.8 with 1N HCl or 1N NaOH before autoclaving. Cultures were incubated at 25 ± 2° C and under continuous light at 450-642 μW . cm^{-2} from a cool, white fluorescent light source (Crompton 40W).

RESULTS

1. *Populus × euramericana*

The results of internodal and nodal explants excised from a 20 to 25 year old hybrid species, *Populus × euramericana* Guinier, and calli derived from the base of *in vitro* raised shoots are given below.

(i) *Response of internodal explants*

Multiple shoots differentiated at the cut ends as well as from the lenticels on the internodal pieces in 20-day-old cultures on MS basal medium (Fig. 1). The number of shoots developed per explant varied from 5 to 15. After five-six

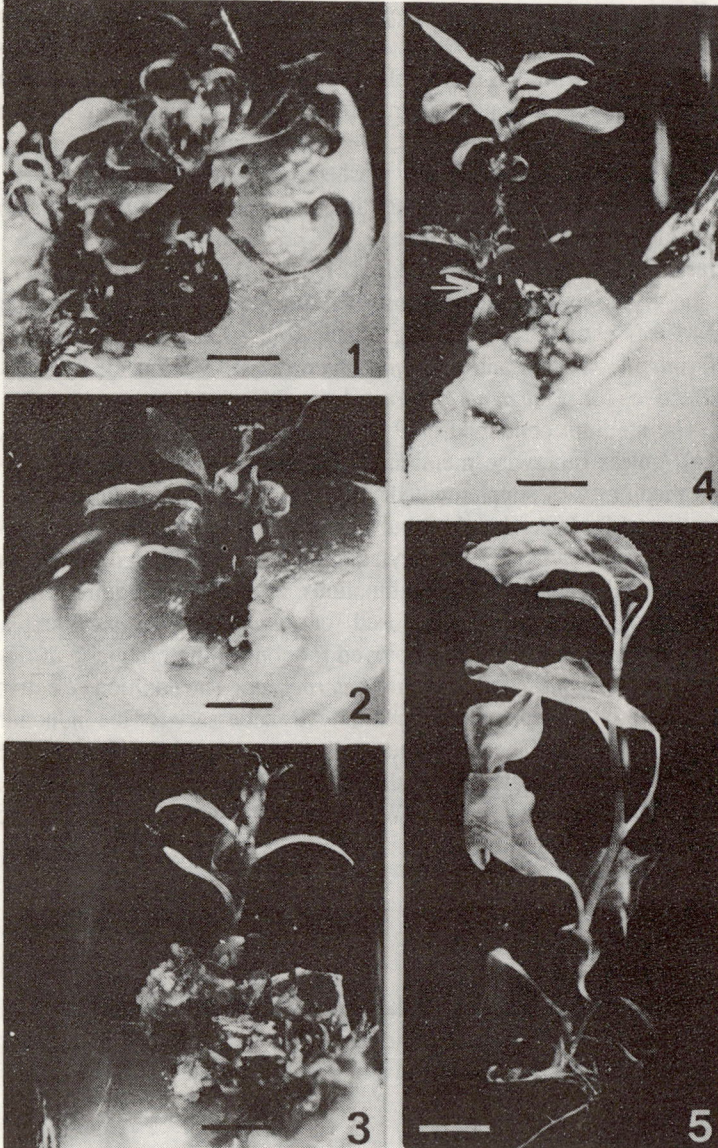

Fig. 1-5

weeks, shoots attaining a length of 2-2.5 cm were transferred individually to media augmented with different auxins (IAA, IBA and NAA) at varying concentrations (0.1-10 μM) for rooting. Explants bearing the stunted shoots were shifted directly to GA_3 (0.1-10 μM) supplemented medium for further elongation of shoots but it did not prove effective.

(ii) *Response of nodal explants*

Like the internodal segments, multiple shoots also developed at the cut ends of explants as well as from the axil of leaves, in 20-day-old cultures on MS basal medium (Fig. 2). Fewer shoots developed in the axils of leaves than at the cut ends of nodal segments. The shoots were excised after attaining a length of 2-2.5 cm.

The excised shoots developed on nodal and internodal explants were subcultured on auxin supplemented media (IAA, IBA, NAA at 0.1-10 μM concentrations). In a 25-day-old culture they formed calli and additional shoots at their basal ends instead of organising roots (Fig. 4). In the second passage, after removing the calli, such shoots were again transferred to different basal media [21, 34, 54] in an attempt to induce roots. However, they neither developed calli nor roots and eventually turned brown. Finally, they withered in 20 days. The shoots did not develop roots even after transferring them to 1/2 or 1/10th strength of MS salts. Some of the shoots were shifted to soil after maintaining them for one week in an auxin supplemented basal medium, but they survived only up to 10 days. Some of the shoots were subcultured on a modified B_5 medium [37] but no roots were induced.

However, when activated charcoal (0.25-2%) was added to 10 μM NAA supplemented MS basal medium, slender, long and branched roots developed at the base of shoots within 15 days (Fig. 5). Maximum frequency of rooting (44%) was obtained at 0.75% AC (Table 1). But the frequency of rooting enhanced significantly to 80% when these shoots were subcultured on the same medium (AC augmented) in 100 ml conical flasks (instead of culture tubes) wherein the root growth also improved.

Figures 1-5:   *In vitro* morphogenic response of internodal, nodal and callus explants of *Populus × euramericana*.

1, 2.   Differentiation of multiple shoots from internodal and nodal explants, respectively from 20-day-old cultures on MS basal medium. Bar represents 4.5 mm.

3.   Organisation of multiple shoots in 15-day-old callus derived from the base of shoots raised *in vitro* on MS + 0.2 μM BA medium. Bar represents 5 mm.

4.   Formation of callus and additional shoots (arrow) at the base of a shoot regenerated *in vitro* on an auxin-supplemented medium (MS + 10 μM NAA) in a 25-day-old culture. Bar represents 4.5 mm.

5.   Development of roots at the base of a shoot raised *in vitro* on MS + 10 μM NAA + 0.75% AC medium in 15-day-old subculture. Bar represents 1 cm.

Table 1: Response of *in vitro* raised shoots to activated charcoal on MS + 10 μM NAA medium in culture tubes after 30 days of subculture.

Activated charcoal (%)	Number of explants subcultured	Number of shoots organising roots	Percentage of shoots forming roots*
0	48	0	0ᶜ
0.25	44	0	0ᶜ
0.50	40	0	0ᶜ
0.75	45	20	44.0ᵃ
1.00	46	6	13.2ᵇ
1.25	40	4	10.0ᵇ
1.50	44	0	0ᶜ
1.75	40	0	0ᶜ
2.00	40	0	0ᶜ

* Values in the last column followed by the same superscript are statistically not significant at the 5% level, as checked by employing the chi-square test.

(iii) *Response of calli*

The calli formed at the base of shoots were compact and of three different colours, white, light green and dark green. They (approximately 0.5 cm in diameter) were subcultured on the same medium (MS +10 μM NAA). Within a month they had proliferated to 2.5 cm in diameter. However, when the calli were subcultured on a medium supplemented with varying concentrations of BA (0.2-20 μM), green nodular structures differentiated within one week. In 15-day-old cultures, these nodular structures developed into numerous shoots at lower concentrations of BA (0.2-0.3 μM; Fig. 3). At higher concentrations (5-20 μM), these nodules failed to develop into shoots. The so-developed shoots, when transferred to a root-inducing medium, formed good root systems. In this way, 20-30 shoots per callus culture were obtained at lower concentrations of BA.

(iv) *Effect of various physical factors*

The effect of various factors, such as position of explant on trees, size of explant, light intensity and seasonal variation on morphogenic response was also assessed.

(a) *Effect of explant location on tree and explant size*: The fourth to seventh youngest internodes and third to eighth nodes from the tip were excised from young branches. However, segments beyond the seventh node and internode showed no morphogenic response. The size of the internodes varied from 0.5 to 2 cm (length) and 0.2 to 0.5 cm (diameter) and of the nodes from 0.2 to 1 cm (length) and 0.2 to 0.8 cm (diameter). They were inoculated on MS basal medium. Internodal explants (1.0 × 0.5 cm) and nodal explants (0.8 × 0.7 cm) showed the best response in terms of shoot differentiation. However, with a decrease in thickness and increase in length of the explants, the response decreased considerably.

(b) *Effect of light intensity*: Both sets of cultured explants were incubated simultaneously under different light intensities such as: (i) total darkness, (ii) diffused light (164 μW . cm^{-2}) and (iii) bright light (492 μW . cm^{-2}). The response of cultures incubated under bright light was statistically higher than that of explants maintained in diffused light and under dark conditions (Table 2).

Table 2. Effect of different light intensities (bright light = 492 μW . cm^{-2}, diffused light = 164 μW . cm^{-2}) and total darkness on morphogenic response of internodal and nodal explants on MS medium after 40 days of incubation.

Explant	Light intensity (μW . cm^{-2})	No. of explants	Percentage of explants developing		
			Callus	Shoots*	Roots
Internode	Dark	72	24.5	8.3[c]	0
	164	72	6.1	32.0[b]	0
	492	146	17.1	52.7[a]	0
Node	Dark	72	34.7	6.3[c]	0
	164	72	44.4	27.7[b]	0
	492	72	27.7	55.5[a]	0

* Numerals followed by the same superscript are statistically not significant at the 5% level, as checked by employing the chi-square test.

(c) *Seasonal variation*: The morphogenic behaviour of nodal and internodal explants varied considerably during different months of an year. The maximum response of explants (90-95%) was seen from March to May, 65-75% from June to August, and 50-58% in September-October. The response declined markedly to 5-0% during November and February. In winter (November to February), the explants turned brown and did not develop shoots even after one month of inoculation. They remained non-responsive even after augmenting the medium with various growth hormones. Probably, some secondary metabolites such as polyphenols are formed in the tissues, which inhibit differentiation.

2. *Acacia* spp.

Acacia auriculiformis and *A. nilotica* have been micropropagated using juvenile explants. Among the different explants tried, multiple shoots could be induced only in axillary buds while the other explants, such as leaf, cotyledon and hypocotyl did not differentiate shoots even after augmenting the medium with several hormones. However, all the explants, except axillary buds, developed roots on the differentiation medium. Multiple shoots originated from axillary buds within 20-25 days on B_5 + CW (5 or 10%) + 1 μM BA medium. Their frequency was higher in medium containing 5% CW.

Initially, these shoots did not develop roots, when subcultured on the same medium or B_5 alone. Transfer of parent explants, along with multiple shoots to a medium containing IAA (0.1-1 μM) helped root initiation in about two months. Its frequency was higher in NAA (0.1 or 1 μM) supplemented medium but the roots were small and thin. Subsequently, the individual shoots of *A. auriculiformis,* on subculture in a medium containing IAA or NAA (0.1 or 1 μM) developed healthy roots without callusing within 50-60 days [32]. Such plantlets have been maintained on the basal medium in culture tubes for further growth. Efforts are being made to increase the number of plantlets for field trials.

3. *Albizia* spp.

This genus had a definite edge over others; in this taxon, besides organogenesis, regeneration was also achieved through somatic embryogenesis [48-50].

(i) *Organogenesis*:

Explants of *in vitro* raised seedlings, such as cotyledon, hypocotyl and root of *A. amara, A. lucida, A. procera* and *A. richardiana,* were tried in order to assess their morphogenic response. In all the species, 1 cm long hypocotyl segments and root explants of *A. amara* and *A. lucida* differentiated shoot buds on B_5 basal medium after 10-15 days of inoculation. However, cotyledonary explants required BA to elicit the same response. A comparative study was made with hypocotyl explants of *A. amara, A. lucida* and *A. richardiana* under varying levels of auxins and cytokinins added to the medium individually. All these species followed almost similar trends for root and shoot differentiation. Auxins, in general, increased production of white friable callus and roots but BA enhanced shoot bud differentiation as well as development of bright green and compact calli. BA at 10 μM concentration was optimal for caulogenesis in all the four species. Initially, rooting of excised shoots was difficult in all the species of *Albizia*, however, a two-step treatment proved helpful. In the first step, the *in vitro* excised shoots along with the mother explants were cultured in B_5 + 1 μM IAA medium for 30 days and in the second step, only the excised shoots were transferred to the same medium, in which 30-40% of them developed roots. The *in vitro* grown plantlets of *A. amara* and *A. richardiana* were transferred to vermiculite and finally to the field, where they are thriving well.

(ii) *Somatic embryogenesis*

Somatic embryogenesis was induced in callus cultures derived from hypocotyl explants of *A. richardiana*. Just by reducing the size of the explants from 10 to 1 mm, the frequency as well as amount of calli per explant enhanced significantly. Interestingly, differentiation of the shoot buds did not take place on small explants (0.1 cm) either on B_5 basal alone or at any concentration of BA tried. By augmenting the B_5 and MS basal media with BA, the morphogenic response of green callus cultures and 1 cm long hypocotyl segments was significantly increased. The frequency of embryos varied from 2 to 20 per embryogenic callus mass. The somatic embryos were typically bipolar with distinct radicular and plumular ends (49). Typical somatic embryos, when subcultured on MS medium containing 0.06 M sucrose, produced bright green and compact calli at the plumular end, which again developed a fresh crop of somatic embryos. Only a few of these grew into complete plantlets. However, the isolated embryos developed into normal plantlets on Knop's medium. Such plants were successfully transferred to vermiculite and have been growing in it for the past few weeks [50].

4. *Leucaena leucocephala*

Epicotyl, cotyledon, hypocotyl and root segments excised from *in vitro* raised 4-6 cm long, 12- to 15-day-old seedlings were cultured on B_5 and MS media supplemented with various concentrations of auxins (IAA, NAA and 2, 4-D) and cytokinins (BA and Kn) either alone or in various combinations. Explants of epicotyl, cotyledon, hypocotyl and root produced calli as well as roots. Shoot buds differentiated from cotyledons, reared on B_5 + BA (4.44, 8.88 or 17.76 µM). Axillary buds developed shoots on Nitsch's medium (36) containing either 10 µM IAA or 1 µM NAA + 10 µM Kn but multiple shoots were organised on NB + 10 µM BA. These shoots, on prolonged incubation, differentiated roots at the base and thus formed plantlets (22). Addition of various concentrations of AC, CW, L-cysteine-HCl and PVP individually to B_5 medium resulted either in callusing or callusing followed by rooting.

Shoots apices (2-3 mm long) containing 3 or 4 leaf primordia, excised from 5-year-old trees were cultured on B_5, MS and NB basal media alone or the same supplemented with various hormones. On MS + 13.33 µM BA + 0.26 µM NAA, they produced multiple shoots. The differentiation of shoots was preceded by callusing. The shoot buds, when subcultured on the same medium, proliferated and increased in number. These shoots were further subcultured on MS + 5.71 µM NAA medium and, after about 4 weeks, a few shoots developed roots at their cut ends.

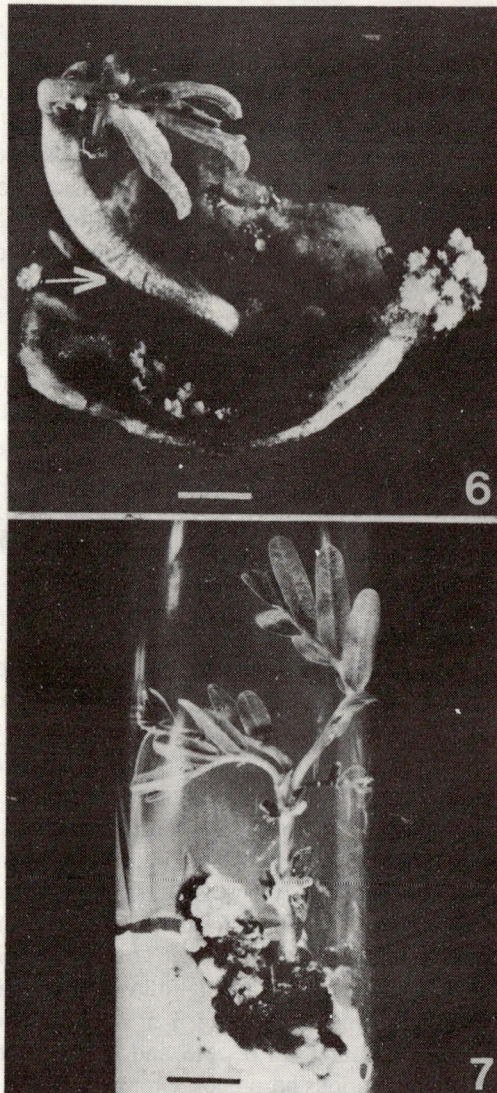

Figs. 6-7: *In vitro* morphogenic response of leaflet explants of *Sesbania macrocarpa*.
6. Differentiation of shoot from a pinna on B_5 + 10 μM BA + 0.1 μM NAA + 1.25 μM PVP medium in a 30-day-old culture on a Petri plate. Bar represents 5 mm.
7. An elongated shoot developed from shoot bud in a test tube on the same medium as mentioned for Figure 6. Bar represents 7.5 mm.

5. *Sesbania macrocarpa*

Micropropagation in this taxon was achieved using explants from *in vitro* raised seedlings as well as field-grown plants.

(i) *Response of juvenile explants*

Approximately 1 cm long segments of cotyledon, hypocotyl and root were inoculated either on B_5 alone or on the same medium supplemented with various growth regulators. The hypocotyl explants differentiated small nodular outgrowths within 10-15 days even on B_5 basal medium, which subsequently developed into shoots. However, for cotyledonary explants, cytokinin was necessary for differentiation, whereas the root segments developed only small calli on all the media tried. BA (10 µM) was optimum for differentiation of shoots [24].

Shoots could also be induced in leaflet explants if B_5 medium was augmented with NAA and PVP in addition to BA. Pieces of pinnae 6-8 mm long, excised from 14 to 16-day-old *in vitro* raised seedlings, differentiated shoot buds on B_5 + 10 µM BA+0.1 µM NAA+1.25 µM PVP medium within 30-40 days (Fig. 6). Initially, the explants were subcultured in 85 mm disposable Petri plates and, after induction of shoots, the explants were transferred to test tubes for further elongation (Fig. 7). Thus, 15-20 shoots per explant could be derived from each pinna explant.

Such shoots, after attaining a height of 2-3 cm, were excised and subcultured on B_5 medium supplemented with auxins. Of the different auxins tried, IBA at 10 µM proved best for rhizogenesis. Initial dark incubation for 7-10 days was necessary for the initiation of roots, whereas for further elongation and growth of plantlets, light was required. Plantlets with an established root system were transferred to pots containing vermiculite, which were kept in a culture room for 10-15 days for exposure to high humidity. These were then transferred to pots containing soil and maintained in a glasshouse for another 15-20 days. Finally, they were transferred to the field where they flowered and set seeds normally [24].

(ii) *Response of field-grown plants*

After ten months of transfer to the field, plants with a faster growth rate were selected for further experiments. Approximately 1 cm long nodal and internodal segments from 5 mm thick branches were excised and inoculated on B_5 alone or BA augmented media. Only the nodal segments differentiated multiple shoots after 15-20 days of inoculation in 1 µM and 10 µM BA augmented media. Maximum shoots (8-10 per explant) differentiated from nodes but remained stunted in a medium containing BA. Further elongation of these shoots was achieved by transferring them to 10 µM GA_3 augmented B_5 medium. The 2-3 cm long shoots were excised and subcultured on B_5 with 10 µM IBA for rooting. Complete plantlets were transferred to soil, following the procedure described earlier [25].

DISCUSSION AND CONCLUSIONS

During the course of the present investigations it was realised that the morphogenic response of tissues is greatly influenced by the composition of the nutrient media, growth adjuvants, size and location of explant on the seedling/tree, type and source of explants, quality and quantity of light, etc. However, the crucial factor affecting morphogenesis remains the delicate balance between the endogenous hormone levels in the tissues and the exogenous, supplied through the culture medium.

In almost all the woody taxa investigated thus far, cytokinins have played a major role in shoot differentiation. BA was necessary for development of shoots in cotyledons and leaves of *Albizia* spp., *Leucaena leucocephala,* and callus of *Populus* × *euramericana* and *Sesbania macrocarpa*, whereas the hypocotyl segments of *Albizia* spp. [51] and *Sesbania macrocarpa* [24], internodal segments of hybrid *Populus* [13], as well as nodal and internodal segments of *Populus* × *euramericana* developed shoots even on the basal medium. This suggested the existence of a sufficient endogenous level of cytokinin in the hypocotyl as well as nodal and internodal segments, but significantly less in the cotyledons and leaves. Lower concentrations of BA (0.2 to 0.3 µM) were indispensable for differentiation of shoots in callus of the presently investigated species of *Populus*, similar to its other species, e.g., *Populus tremuloides* [6, 37, 55, 56], *Populus nigra* [53], and 12 other species of *Populus* [44]. Higher concentrations of BA (5-20 µM) produced only numerous stunted shoot buds in *Populus* calli. Similar observations have been recorded earlier in various other species of *Populus* [44]. The requirement for a low concentration of cytokinins further suggests that callus is capable of synthesising cytokinins in sufficient amounts, a view expressed earlier for *Pinus sylvestris* [23].

The addition of auxin to the medium was necessary for induction of roots. The role of auxins has also been reported earlier by various workers [9, 35, 40, 45, 52]. Besides, a two-step method was found to be effective in inducing roots in *Albizia* spp. [51] and *Acaacia auriculiformis* [32]. However, in the hybrid *Populus* investigated in our laboratory, roots could be induced only by adding AC to the medium containing auxin. The probable role of AC is the adsorption of toxic substances which inhibit induction of roots. This view accords with that of Fridborg et al. [20]. A decline in the rooting percentage beyond 0.75% of AC, further suggested that at higher levels, it might be adsorbing even some of the essential components, in addition to the inhibitory compounds, as also recorded for *Pinus sylvestris* [23].

The size of explants affects morphogenic response in various taxa. Interesting observations have been made for *Albizia richardiana;* small explants (1 mm long) produced only calli but the 10 mm long explants developed only shoots [50, 51]. Differential morphogenic responses with different sizes of explants have also been noticed in poplars [44]. In the

plants have also been noticed in poplars [44]. In the presently investigated hybrid of *Populus*, too, the internodal explants of 1×0.5 cm yielded the best response in terms of shoot differentiation, whereas with a decrease in thickness and increase in length of the explants, the response decreased considerably.

The seasons of the year influence the morphogenic potential of the *Populus* tree to a great extent, as recorded earlier for *Chrysanthemum* [39] and *Pinus sylvestris* [23]. Due to the accumulation of polyphenols, the tissues turned brown and failed to respond even after fortifying the medium with various hormones. Though the browning can be checked to some extent by different anti-oxidants, such as tyrosine [42], PVP [2] L-cysteine-HCl [57], ascorbic acid [23, 27] and citric acid [23], it is not possible to prevent synthesis of polyphenols [3, 23].

The list of woody plants developed through tissue culture has increased during the last decade [4, 10, 15]. But only a few of them can be exploited commercially or for forestry programmes as their large-scale acclimatisation to field conditions has not yet been standardised. The basic requirement is to make available the standard protocols for hardening, prior to the transfer of plantlets to soil, and of course, the resistance of plants to pathogens and herbicides, etc. needs further investigations. Refinement in the available techniques can no doubt be made so that the plants could be transferred to the field successfully [7, 18, 26, 43]. Attempts to select pathogen- and herbicide-tolerant plants through somaclonal variations will hopefully be rewarding [30].

More recently, the technique of genetic engineering has also opened a new vista for improving the woody plants by incorporating the desired gene constructs into the tree genome. So far, genetic transformations have been reported for just a few taxa, e.g., *Betula papyrifera* and *Populus tremuloides* [31], *Populus* hybrid NC-5339 (*Populus alba* × *grandidentata*) [19], *Juglans regia* [12], etc., but this aspect is still in its infancy and needs more attention.

Regeneration through somatic embryogenesis is yet another potential mode of micropropagation [4, 16] since it is possible through somatic embryos to develop artificial 'seeds' which can be stored safely [46]. The possibility of long-term storage of embryos through cryopreservation also exists. This aspect is currently drawing more attention.

Needless to say, in view of the urgent need for increased production of biofuel energy and other forest-based products, *in vitro* micropropagation has far-reaching implications in mass production and tree improvement. Looking at the pace of the progress made in this area, it is evident that tissue culture technology is more than competent to achieve the desired targets within a reasonable period, if used judiciously.

Acknowledgements
We are thankful to Dr. Kanan Nanda, Dr. Virendra Gautam (*Leucaena leucocephala*), Mrs. Aradhana Mittal (*Acacia auriculiformis*), Dr. Uttar Kumar

Tomar (*Albizia* spp.) and Mrs. Seema Dhir (*Sesbania macrocarpa*) for their participation in research. These investigations were financed in part by grant No. FG-IN-619 (IN-FS-95) made to SCG by the United States Department of Agriculture under the Co-operative Agricultural Research Programme. VA is grateful to the Council of Scientific & Industrial Research, New Delhi, for the award of a Research Associateship and to the University Grants Commission for her appointment as Research Scientist 'A'.

LITERATURE CITED

1. Ahuja, M.R. 1987. *In vitro* propagation of poplar and aspen. In: *Cell and Tissue Culture in Forestry* edited by J.M. Bonga and D.J. Durzan. Martinus Nijhoff, The Netherlands, pp. 207-223.
2. Andersen, R.A. and J.A. Sowers. 1968. Optimum conditions for bonding of plant phenols to insoluble polyvinylpyrrolidone. *Phytochemistry*, **7**: 293-301.
3. Babbar, S.B. and S.C. Gupta. 1982. Promotory effect of polyvinylpyrrolidone and L-cysteine-HCl on pollen plantlet production in anther cultures of *Datura metel*. *Z. Pflanzenphysiol.*, **106**: 459-464.
4. Bajaj, Y.P.S. 1986. Biotechnology of tree improvement for rapid propagation and biomass energy production. In: *Biotechnology in Agriculture and Forestry* edited by Y.P.S. Bajaj. Springer-Verlag, Berlin, vol. I., pp. 1-23.
5. Barghchi, M. 1988. Micropropagation of *Alnus cordata* Loisel. *Pl. Cell Tissue Organ Cult.*, **15**: 233-244.
6. Barocka, K.H., M. Baus, E. Lontke and F. Sievert. 1985. Tissue culture as a tool for *in vitro* mass-propagation of aspen. *Z. Pflzücht.*, **94**: 340-343.
7. Bonga, J.M. 1977. Applications of tissue culture in forestry. In: *Applied and Fundamental Aspects of Plant Cell, Tissue and Organ Culture* edited by J. Reinert and Y.P.S. Bajaj. Springer Verlag, Berlin, pp. 93-108.
8. Bonga, J.M. 1981. Vegetative propagation of mature trees by tissue culture. In: *Tissue Culture of Economically Important Plants* edited by A.N. Rao. COSTED and ANBS, Singapore, pp. 191-196.
9. Bonga, J.M. 1985. Vegetative propagation in relation to juvenility, maturity and rejuvenation. In: *Tissue Culture in Forestry* edited by J.M. Bonga and D.J. Durzan. Martinus Nijhoff, The Netherlands, pp, 387-412.
10. Brown, C.L. and H.E. Sommer. 1985. Vegetative propagation of dicotyledonous trees. In: *Tissue Culture in Forestry* edited by J.M. Bonga and D.J. Durzan. Martinus Nijhoff, The Netherlands, pp. 109-149.
11. Cheema, G.S. and D.P. Sharma. 1983. *In vitro* propagation of apple. In: *Plant Cell Culture in Crop Improvement* edited by S.K. Sen and K.L. Giles. Plenum Press, New York, pp. 309-317.
12. Dandekar A.M., L.A. Martin and G. McGranahan. 1988. Genetic transformation and foreign gene expression in walnut tissue. *J. Am. Soc. Hort. Sci.*, **113**: 945-949.
13. Douglas, G.C. 1985. Formation of adventitious buds in stem internodes of *Populus* hybrid TT 32 cultured *in vitro*: effects of sucrose, zeatin, IAA and ABA. *J. Pl. Physiol.*, **121**: 225-231.
14. Douglas, G.C. 1989. Poplar (*Populus* spp.). In: *Biotechnology in Agriculture and Forestry* edited by Y.P.S. Bajaj. Springer-Verlag, Berlin, vol. 5, pp. 300-319.
15. Dunstan, D.I. and T.A. Thorpe. 1986. Regeneration in forest trees. In: *Cell Culture and Somatic Cell Genetics of Plants* edited by I.K. Vasil. Academic Press Inc., New York, vol. 3, pp. 223-241.

16. Durzan, D. 1982. Somatic embryogenesis and sphaeroblasts in conifer cell suspension. In: *Plant Tissue Culture* edited by A. Fujiwara. Maruzen, Tokyo, pp. 113-114.

17. Durzan, D.J. 1985. Cell and tissue culture in forest industry. In: *Tissue Culture in Forestry* edited by J.M. Bonga and D.J. Durzan. Martinus Nijhoff, The Netherlands, pp. 36-71.

18. Farnum, P., R. Timmis and L. Kulp. 1983. Biotechnology of forest yield. *Science*, **219**: 694-702.

19. Fillatti, J.J., J. Sellmer, B. McCown, B.E. Haissig and L. Comai. 1987. *Agrobacterium* mediated transformation and regeneration of *Populus*. *Mol. Gen. Genet.*, **206**: 192-199.

20. Fridborg, G., M. Pedersen, L.E. Landström and T. Eriksson. 1978. The effect of activated charcoal on tissue cultures: adsorption of metabolites inhibiting morphogenesis. *Physiol. Plant.*, **43**: 104-106.

21. Gamborg, O.L., R.A. Miller and K.Ojima. 1968. Nutrient requirements of suspension cultures of soybean root cells. *Exp. Cell Res.*, **50**: 151-158.

22. Gautam, V.K., K. Nanda and S.C. Gupta. 1985. Morphogenic responses of a leguminous tree, *Leucaena leucocephala*. *Natn. Seminar on Reproduction Biology of Plants Including Endangered Species, Meerut*, pp. 32-33.

23. Hohtola, A. 1988. Seasonal changes in explant viability and contamination of tissue cultures from mature scot pine. *Pl. Cell Tissue Organ Cult.*, **15**: 211-222.

24. Kapoor, S. and S.C. Gupta. 1986. Rapid *in vitro* differentiation of *Sesbania bispinosa* plants — a leguminous shrub. *Pl. Cell Tissue Organ Cult.*, **7**: 263-268.

25. Kapoor, S. and S.C. Gupta. 1987. Micropropagation of a leguminous shrub — *Sesbania bispinosa*. *2nd A. Conf. Int. Pl. Biotech. Network, Bangκok, Thailand*, Abstr. No. 21.

26. Karnosky, D.F. 1981. Potential for forest tree improvement via tissue culture. *Bio-Science*, **31**: 114-120.

27. Larson, R.A. 1988. The antioxidants of higher plants. *Phytochemistry* **27**: 969-978.

28. Mehra, P.N. and G.S. Cheema. 1980. Clonal multiplication *in vitro* of Himalayan poplar (*Populus ciliata*). *Phytomorphology*, **30**: 336-343.

29. Mehra, P.N. and G.S. Cheema. 1985. Differential response of male and female Himalayan poplar (*Populus ciliata*) and *P. alba in vitro*. *Phytomorphology*, **35**: 151-154.

30. Michler, C. and E. Bauer. 1987. Somatic embryogenesis in plant cell cultures of *Populus*. In vitro, 23 Abstr. No. 140.

31. Minocha, S.C., E.W. Noh and A.P. Kausch. 1986. Tissue culture and genetic transformation in *Betula papyrifera* and *Populus tremuloides*. In: *Res. and Dev. Conf. Tech. Park/Atlanta*, pp. 89-92.

32. Mittal, A., R. Agarwal and S.C. Gupta. 1989. *In vitro* development of plantlets from axillary buds of *Acacia auriculiformis* — a leguminous tree. *Pl. Cell Tissue Organ Cult.*, **19**: 65-70.

33. Muralidharan, E.M. and A.F. Mascarenhas. 1987. *In vitro* platlets formation by organogenesis in *E. camaldulensis* and by somatic embryogenesis in *Eucalyptus citriodora*. *Pl. Cell Reports*, **6**: 256-259.

34. Murashige, T. and F. Skoog. 1962. A revised medium for rapid growth and bioassay with tobacco tissue cultures. *Physiol. Plant.*, **15**: 473-497.

35. Németh, G. 1986. Induction of rooting. In: *Biotechnology in Agriculture and Forestry* edited by Y.P.S. Bajaj. Springer Verlag, Berlin, vol. 1, Trees 1. pp. 49-64.

36. Nitsch, J.P. 1969. Experimental androgenesis in *Nicotiana*. *Phytomorphology*, **19**: 389-404.

37. Noh, E.W. and S.C. Minocha. 1986. High efficiency shoot regeneration from callus of quacking aspen (*Populus tremuloides*). *Pl. Cell Reports*, **5**: 464-467.

38. Park, Y.G. and S.H.Son. 1988. Regeneration of plantlets from cell suspension culture

derived callus of white poplar (*Populus alba* L.). *Pl. Cell Reports*, **7**: 567-570.

39. Prasad, R.N. and H.C. Chaturvedi. 1988. Effect of season of collection of explants on micropropagation of *Chrysanthemum morifolium*. *Biol. Plant.*, **30**: 20-24.

40. Rahman, M.A. 1988. Effect of nutrients and IBA on the *in vitro* rooting and *ex vitro* establishment of jack fruit (*Artocarpus heterophyllus* Lam.). *Bangladesh J. Bot.*, **17**: 105-110.

41. Rao, P.V.L. and D.N. De. 1987. Tissue culture propagation of tree legume *Albizia lebbeck* (L.) Benth. *Pl. Sci.*, **51**: 263-267.

42. Reinert, J. and P.R. White. 1956. The cultivation *in vitro* of tumor tissues and normal tissues of *Picea glauca*. *Physiol. Plant.*, **9**: 177-189.

43. Reinert, J. and Y.P.S. Bajaj. 1977. *Applied and Fundamental Aspects of Plant Cell, Tissue and Organ Culture.* Springer-Verlag, Berlin.

44. Rutledge, C.B. and G.S. Douglas. 1988. Culture of meristem tips and micropropagation of 12 commercial clones of poplars *in vitro*. *Physiol. Plant.*, **72**: 361-373.

45. Scott, T.K. 1972. Auxins and roots. *A. Rev. Pl. Physiol.*, **23**: 235-258.

46. Sharp, W.R., D.A. Evans and M.P. Sondahl. 1982. Application of somatic embryogenesis to crop improvement. In: *Plant Tissue Culture* edited by A. Fujiwara. Maruzen, Tokyo, pp. 759-762.

47. Sita, G.L., S. Chattopadhyay and D.H. Tejavathi. 1986. Plant regeneration from shoot callus of rosewood (*Dalbergia latifolia* Roxb.). *Pl. Cell Reports*, **5**: 266-268.

48. Tomar, U.K. and S.C. Gupta. 1986. Organogenesis and somatic embryogenesis in leguminous trees — *Albizia* spp. *6th Int. Congr. Pl. Tissue Cell Cult.*, Univ. Minnesota, Minneapolis, USA, p. 41.

49. Tomar, U.K. and S.C. Gupta. 1987. High frequency somatic embryogenesis in a leguminous tree, *Albizia richardiana* King. *2nd A. Conf. Int. Pl. Biotech. Network, Bangkok, Thailand*, Abstr. No. 55.

50. Tomar, U.K. and S.C. Gupta. 1988a. Somatic embryogenesis and organogenesis in callus cultures of a tree legume — *Albizia richardiana* King. *Pl. Cell Reports*, **7**: 70-73.

51. Tomar, U.K. and S.C. Gupta, 1988b. *In vitro* plant regeneration of leguminous trees — *Albizia* spp. *Pl. Cell Reports*, **7**: 385-388.

52. Torrey, J.G. 1976. Root hormones and plant growth. *A. Rev. Pl. Physiol.*, **27**: 435-459.

53. Venverloo, C.J. 1973. The formation of adventitious organs. I. Cytokinin induced formation of leaves and shoots in callus cultures of *Populus nigra* L. "*Italica*". *Acta bot. neerl.*, **22**: 390-398.

54. White, P.R. 1943. *Handbook of Plant Tissue Culture.* Jaques Cattell, Lancaster.

55. Winton, L. 1970. Shoot and tree production from aspen tissue cultures. *Am. J. Bot.*. **57**: 904-909.

56. Wolter, K.E. 1968. Root and shoot initiation in aspen callus cultures. *Nature*, **219**: 509-510.

57. Yeoman, M.M. 1973. Tissue (callus) culture techniques. In: *Plant Tissue and Cell Culture* edited by H.E. Street. Blackwell, Oxford, pp. 31-58.

5

Productivity of Tea (*Camellia sinensis* (L.)) in Relation to Other Tropical Perennial Crops

S. Kulasegaram and G. Wadasinghe

Tea Research Institute of Sri Lanka, Talawakelle, Sri Lanka

ABSTRACT

High yields and dry matter production rates have been obtained for many tropical perennial crops but the productivity of tea appears to be disappointingly low in comparison. This is attributed to the inefficient conversion of intercepted radiation to dry matter and the very low harvest index compared to the other plantation crops, which also have low harvest indices compared to many temperate annual crops. The low yield is also ascribed to sink limitation because the method of harvesting removes the small number of individual sinks as they are about to grow most rapidly; thus low yield is not due to a shortage of potential sinks.

Here hormonal interactions between shoots and roots are likely to determine the actual number of growing shoots and consequently the influence the soil factors are likely to determine the yield. Seasonal variation in yield is known to be due to the number and weight of shoots per unit area of ground and the rate of growth of shoots which is controlled by air temperature and potential transpiration rate.

In this context, analyses of light interception and penetrtation, canopy depth and leaf area index, photosynthetic rates, leaf pose in relation to leaf temperature, leaf longevity, reduction of maintenance of respiration rates, cultural and bush management practices etc. of clones differing in yield, should provide useful pointers to improve productivity.

Limitations to high productivity are also highlighted in relation to other tropical perennial crops and suggestions are made to improve net productivity.

INTRODUCTION

Perennial crops occupy an important place in tropical agriculture not only because of the range of foods, beverages and industrial raw materials that they produce, but also because their perennial habit is particularly suited to the wet tropics where minimum cultivation is desirable to prevent rapid weathering and erosion and the ambient temperatures permit year-round growth.

Some crops are grown as isolated bushes in mixed stands, e.g. coffeee and pepper while others are grown as plantation crops in closed stands with a continuous leaf canopy, e.g. tea, rubber and oil palm, and their produce also varies from those of vegetative crops such as grasslands, forests, root crops (cassava), tea, and *Hevea*, and from those of reproductive crops such as coconuts, oil palm etc. This paper reviews the yield and productivity of tea in relation to several other tropical perennial C_3 species for which adequate data on total dry matter production are available and also estimates their potential yields. Some limitations to productivity are identified and suggestions made for improving productivity.

CROPS AND THEIR IMPORTANCE

The crops considered are listed in Table 1 together with their major products and typical yields achieved with good management in favourable environments. Perennial crops, in contrast to those of annuals, have an immature period before harvesting commences and thus, after reaching a peak, yields may decline gradually as the trees age. The immature period and economic life typical of these crops are also given in Table 1. Notwithstanding these limitations, the yields of starch, oil and protein of most crops compare favourably with those reported for cereals, annual oilseeds and leguminous crops. When

Table 1: Details of the crops reviewed

Crop	Main Product	Typical Yield (years)	Economic Life	Immature Period (years)
		(t ha^{-1}a^{-1})		
Oil palm	oil	5–6	25–30	2.5
Coconut	copra	2–3	up to 70	2.7–7
Rubber	rubber	2	25–30	5
Cocoa	cocoa beans	1.5–2.5	30–70	1.5
Cassava	starch	10	0.5–2	–
Sago	starch	10	unknown	8
Leucaena	forage/	10–15	unknown	0.5
	protein	2–3		
Tea	tea	1.5–2.5	30–50	2

comparing the yields of these crops a distinction must be made between the harvested and the economic products (Tables 2, 3, 4) [6].

Table 2: Constituents of harvested products

Crop	Harvested product	Major economic product	Other harvested material	Uses for other material
Oil palm	fruit branches	mesocarp oil and kernel oil	branch, stalks, shells, and fibre	mulch/ fertiliser or fuel
Coconut	fruits	endosperm (copra)	husks (coir) and shells	coconut matting fuel/ activated charcoal
Rubber	latex	rubber	latex sugars	waste
Cocoa	fruits	seeds (beans)	pod husks	waste
Sago	trunk	starch	bark	waste
Cassava	roots	starch	cortex	waste
Leucaena	leaves and stems	leaves	wood stems	waste
Tea	tender shoots	manufactured tea	coarse stems and prunings	mulch/ fertiliser

Table 3: Composition of harvested products
(for trees with best partitioning ratios)

Crop	Percentage of Total Dry Weight in Economic Products			Waste Material		
	Lipid	Protein	Carbohydrates	Rubber	Lignin	Fibre
Oil palm	65	-	-	-	10	25
Coconut	29	3	10	-	19	39
Rubber	-	-	-	90	-	10
Cocoa	30	4	12	-	-	54
Cassava	-	-	89	-	-	11
Sago	-	-	80	-	-	20
Leucaena	-	22	57	-	-	21

Composition of tea is given in a separate table.

The main product of the oil palm (*Elaeis gumeensis*) is palm oil obtained from the mesocarp. Palm kernel oil, which has a different fatty acid composition, is produced in smaller quantities, and kernel cake is used in animal feeds after extraction of the oil.

Table 4: Approximate general analysis of black tea

Component	Fresh flush	Black tea	Black tea brew
Proteins	15	15	trace
Fibre	30	30	0
Pigments	5	5	trace
Caffeine	4	4	3.2
Polyphenols, simple	30	5	4.5
Polyphenols, oxidized	0	25	15
Amino acids	4	4	3.5
Ash	5	5	4.5
Carbohydrates	7	7	4
Volatile compounds	0.01	0.01	0.01

The coconut (*Cocos nucifera*) has numerous uses. In plantations the primary product is copra, the solid part of the endosperm. Coconut oil can be extracted from this and the residue used as animal feed.

The rubber tree (*Hevea brasiliensis*) is tapped by removing parings of bark from the trunk and collecting the latex which flows from the cut. Latex continues to flow for a few hours and tapping is usually repeated every second day, although many different tapping systems are used. The dry rubber content of latex ranges from 10 to 50%, being less when tapping is intense.

Cocoa (*Theobroma cacao*) is a small understorey shade tree which bears fruits on the trunk and main branches. Seeds are removed from the pods, fermented, dried and then processed to yield cocoa powder and cocoa butter.

The two starch crops have contrasting habits. In cassava (*Manihot esculenta*) starch is stored in the tuberous roots, which may be harvested after six months in short-season cultivars whereas others may continue to accumulate starch for at least two years. The sago palm (*Metroxylon sagu*) stores starch in the trunk, during a seven- to fifteen-year vegetative phase. A massive terminal inflorescence is then produced and much of the stored starch is remobilised and utilised for flowering and fruiting. Thus, trunks are best harvested just before flowering and trees will then regenerate from basal suckers. Most sago starch is harvested from wild trees although a system of continuous cropping has been developed in Malaysia.

The leguminous tree, *Leucaena leucocephaela*, is attracting attention both as a fast-growing timber tree and as a forage crop grown in coppice or hedge-row systems. The economic product is the protein in leaves and fine stems, but with mechanical harvesting woody stems will also be cut.

Tea (*Camellia sinesis* (L.)) is essentially a beverage crop, but the manufactured products differ widely depending on the manufacturing process. There are black teas, green teas, instant teas, liquid teas and scented teas etc. Tea is grown as an evergreen bush crop and the harvested product generally consists of the tender apical shoots. These shoots are selectively harvested at regular intervals.

PRODUCTIVITY OF PERENNIAL CROPS

The annual dry matter yield of oil palm, coconut, rubber, cocoa and tea are very much comparable to each other but significantly lower than that of cassava and *Leucaena* (Tables 5, 6) Tea shows the lowest crop growth rate among the perennial crops considered here. Being an evergreen perennial crop which covers the entire ground under cultivation, one could expect a very much higher dry matter production in tea compared to other perennial crops.

Table 5: Annual productivity estimates for crops grown with good management in favourable environments

| Crop | Crop growth rate (t ha^{-1}a^{-1}) | Dry matter in | | Yield (t ha^{-1}a^{-1}) |
		Harvested product (%)	Economic product (%)	
Oil palm	29	42	17	5
Coconut	24	30	10	2.5
Rubber	26	9	8	2.0
Cocoa	20	20	10	2.0
Cassava	22	50	40	10
Leucaena	21	50	50	10
Tea	18	16–19	16–19	3–3.5

Table 6: Best yields, crop growth rates and harvest indices (not necessarily from the same source)

| Crop | Yield (t ha^{-1}a^{-1}) | Crop Growth Rate | | Harvest Index, % dry matter |
		Dry matter (t ha^{-1}a^{-1})	Energy (TJ ha^{-1}a^{-1})	
Oil Palm	9.9	41	0.89	61
Coconut	6.3	31	0.70	62
Rubber	5.0	36	0.71	37
Cocoa	4.4	30	0.62	30
Cassava	25	51	0.96	70
Sago	25	53	0.99	76
Leucaena	23	32	0.60	—
Tea	8	23	0.44	30–40

The dry matter yields of C$_3$ vegetative crops such as grasslands, forests and root crops like cassava, growing at 1500 to 2000 m altitude in fertile soils near the equator, are commonly 10 to 20 t ha^{-1} and their total net biomass production may be 25 to 40 t ha $^{-1}$ and $^{-1}$ [3]. By contrast, tea (*Camellia sinensis* (L.)) in the highlands (>2000 m) yields only 1 to 2.5 t ha^{-1} year^{-1} and rarely more than 4.0 t ha^{-1} year^{-1}at lower altitudes, though these are among the highest tea yields in the world. Furthermore, the total net biomass production of tea is said to be only 15 to 18 t ha $^{-1}$ year $^{-1}$ [15, 21, 28]. Tea yields of 1 to 4 t ha $^{-1}$ year $^{-1}$ of dry shoot tips are much less than those of other vegetative crops, such as grasslands and forests growing in similar conditions.

WHY TEA YIELDS ARE SO LOW

The question has often been asked why has tea such low productivity? It is difficult to believe that the productivities of closed canopies of plantation tea are source limited. Hadfield [7] claimed that horizontal leaved teas in Assam had low yields because they have an unfavourable canopy architecture, but they also have few shoot sinks per unit ground area, and even erect leaved China teas have low biomass productivities compared with other crops. Moreover, Squire [24] showed that potential net photosynthetic rates of tea leaves in the field were no less than those of many other crops; he found little evidence that shoot growth rates were limited by assimilate supplies.

Hadfield [10] argued that tea productivity could be severely sink limited because of repeated plucking and Tanton [28] claimed supporting evidence on the grounds that tea yields in Malawi were 38% less when plucked every seven days rather than every 14 days. He argued that 7-day plucking removed shoot tips before they reached their maximum growth rates and greatest sink capacities, but he did not show whether 7-day plucking decreased total biomass production, the proportion partitioned to leaves or the proportion removed as yield. In any case Huxley's [11] original arguments ignored the fact that there seem to be ample potentially active apical meristems on some clones of conventionally plucked tea. At least 50% of the shoot tips can be dormant even in flush periods and the proportion on dormant shoots on different clones is rarely associated with yield [32, 34].

Magambo and Cannel [17] showed that tea's poor biomass productivity can largely be accounted for by plucking itself, which decreases annual biomass production by 30%. Unplucked tea produced 26.3 t ha^{-1} year^{-1} of dry matter which at 2178 m altitude near the equator may be as much as is produced by C_3 grasslands, forests or root crops. The main effect of plucking is to limit the development of woody stems and hence cambial sinks. In the absence of plucking, 64% of the additional dry matter went to stems.

Tea's low yield compared with C_3 vegetative crop such as grasslands cannot be wholly attributed to its low biomass productivity when plucked because plucking decreases total leaf production by only 20%. The important factor limiting tea yield is its small harvest index (11% in tea compared to 30 to 70% in grasslands, forests and root crops, including the storage root). Others have also recorded similar low harvest indices for tea [12, 15, 28]. Only 27% of the annual increment of unplucked bushes goes to leaves compared with 41% in coffee. Like temperate trees, tea allocates about 50% of its dry matter increment to woody stem tissues and if wood yield of unplucked tea is considered, its harvest index would be 51% and its yield would be 13.4 t ha^{-1} year^{-1}. The crop physiology of tea may be more closely compared with that of *Hevea* rubber, which can produce 26 to 36 t ha^{-1} year^{-1} biomass when untapped but produces only 15 to 20 t ha^{-1} year^{-1} when tapped because tapping checks cambial as well as shoot growth. The harvest index of rubber varies from 3 to 20% and latex yields are 1 to 4 t ha^{-1} year^{-1}., very similar to tea [31].

DRY MATTER PRODUCTION AND PRODUCTIVITY OF TEA

The annual total radiation in the humid tropics is usually between 60 to 75 TJ ha^{-1} a^{-1} (1 TJ = 10 ^{12}J) [3]. Photosynthetically active radiation (PAR) represents approximately 50% of this total [19]. Thus the data in Table 4 represents efficiencies of conversion of PAR into dry matter ranging from 1.6% to slightly more than 3%.

The tea plant is an evergreen shrub which is pruned and kept low for convenience in harvesting its young shoots. Quality restraints restrict the size of the shoots which can be harvested to young ones with less than three leaves. In most tea-growing countries in the tropics, irradiance is considered more than adequate for satisfactory growth year-round.

Tea has traditionally been grown under shade trees where irradiance at the surface of the bush is below 200 Wm^{-2} on cloudless days and shade is still considered essential for optimal, photosynthesis in the tea areas of the world where leaf temperature exceeds 35°C for much of the year [8]. The light saturation point for photosynthesis in mature tea exceeds 700 Wm^{-2} [23] and in single leaves the saturation point is approximaely 350 Wm^{-2} [23, 24]. This relatively high level of light suggests that in the absence of limiting factors such as high leaf temperature or poor plant nutrition, tea plants grown without shade fix more carbon dioxide than shaded ones. This, in fact, was shown to be so in shade experiments in Malawi. Yet, in the absence of shade, tea yields are still low compared to other vegetative crops, and it is necessary, to see if yields are limited by the rate of photosynthesis (Table 7). Squire recorded maximum photosynthetic rate of tea of 0.04 mg CO_2 m^{-2} s^{-1} (leaf area) at 350 Wm^{-2} in Malawi [24]. This is comparable to those obtained for tea elsewhere in Japan and Sri Lanka using an infrared gas analysis technique [33]. Squire gave rates of photosynthesis per unit area of unshaded leaf ranging from 9 to 13 g CO_2m^{-2} day^{-1}. In several crops with high leaf area indices (tea has a LA1 of 5 to 6) the total photosynthetic capacity of the crop is about twice that of the calculated rate based on unshaded LA1 of 1 [35]. He thus calculated the total photosynthetic production to be about 51,600 kg ha^{-1} (CH_2O). Since direct measurement of plant mass indicates that tea accumulates 17.50 t ha^{-1} year^{-1} of dry matter of

Table 7: Maximum rates of photosynthesis

Crop	Rate of photosynthesis ($g\ CO_2\ m^{-2}\ h^{-1}$)
Oil palm	3.0
Rubber	2.0
Cocoa	0.7
Sago	1.3
Cassava	3.0
Tea	2.2

which approximately 16.00 t ha^{-1} year^{-1} is estimated as CH_2O, respiration in the bush appears to account for 67% of total photosynthetic production. Hadfield [9] gave a figure of 85% for tea while Corley [5] estimated 80% for oil palm. Comparable figures were obtained for tropical rain forests [1, 13, 20]. Hence dry matter accumulation would benefit from a reduction in respiration rate. Since the harvested crop is only 2.50 t ha^{-1} year^{-1} this indicates that yield is limited either by lack of assimilates due to preferential partitioning to wood and roots or to limiting factors other than CH_2O supply.

Magambo and Othieno [18] showed that groups of tea bushes that were harvested for a year yielded 1.39 t ha^{-1} year^{-1}, while similar bushes which were not harvested yielded 6.58 t ha^{-1} year^{-1} dry matter at the end of the season. It is concluded that yield is limited by management practices since the plant is capable of producing a high yield if the sink is not limited by harvesting (Tables 8, 9). In tea harvesting, most of the shoots are harvested before they reach their maximum rate of absolute growth and the detrimental effect of plucking small shoots is more pronounced when considered in terms of dry weight [14].

Huxley [11] considered the low yield of tea in relation to its potential total dry matter production and concluded that yield was sink limited, i.e., by the number of shoots that grow per unit area of plucking table and the rate at

Table 8: Total dry matter production and its distribution

	Managed Plants	Free Growing Plants
Total dry matter production (t/ha/yr)	6.3 (100)	11.6 (183)
Percentage		
Leaves	14.9	10.4
Stems	52.8	61.0
Roots	32.2	28.6

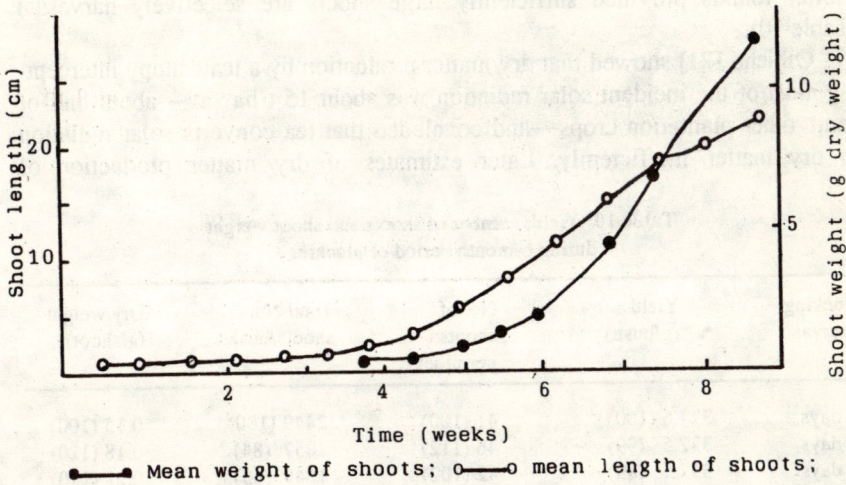

●——● Mean weight of shoots; ○——○ mean length of shoots;

Fig. 1: Growth of tea shoots.

which they extend and fill with dry matter, rather than by the rate at which dry matter is produced and made available to them [11]. Cartwright and Cannel [2] agreed with Huxley [11] that yield was probably sink limited, mainly because the method of harvesting removes individual sinks just as they are about to grow most rapidly and emphasised that sink limitation was the result of a small number of actual sinks (growing shoots) and not a shortage of potential sinks (buds). They proposed that hormonal interactions between roots and shoot (apical dominance) determined the actual number of growing shoots and consequently soil factors were ones most likely to influence yield. Tanton [28] provided evidence that the tea crop, which consists of young shoots, is indeed sink limited and that plucking of the immature shoots, essential for quality tea, is a major factor limiting yields of the crop. This suggests that plucking coarser shoots or lengthening the plucking rounds would result in higher yield. How-

Table 9: Average relative growth rate (RGR), crop growth rate (CGR) and net assimilation rate (NAR)

	Managed Plants	Free Growing Plants
RGR (g/g/wk)	0.0211 (100)	0.0236 (112)
CGR (g/dm²/wk)	0.1167 (100)	0.2409 (206)
NAR (g/dm²/wk)	0.0830 (100)	0.1114 (134)

ever, studies in Sri Lanka [33] have shown that high yields are possible on shorter rounds provided sufficiently large shoots are selectively harvested (Table 10).

Othieno [21] showed that dry matter production by a tea canopy intercepting most of the incident solar radiation was about 15 t ha^{-1} a^{-1}—about half of some other plantation crops—and concluded that tea converts solar radiation to dry matter inefficiently. Later estimates of dry matter production of

Table 10: Yield, number of shoots and shoot weight
during 6-month period of plucking

Plucking interval	Yield (g/bush)	No. of shoots per pluck	Total No. shoots/bush	Dry weight (g/shoot)
3 days	377.6 (100)	41 (100)	2449 (100)	0.15 (100)
4 days	372.5 (99)	46 (112)	2057 (84)	0.18 (120)
5 days	337.5 (89)	42 (102)	1543 (63)	0.21 (140)
6 days	316.0 (84)	52 (127)	1567 (64)	0.20 (133)
7 days	294.5 (78)	60 (146)	1566 (64)	0.19 (127)
8 days	288.6 (76)	56 (137)	1298 (53)	0.22 (147)
9 days	282.6 (75)	46 (120)	980 (40)	0.29 (193)
10 days	260.2 (69)	60 (146)	1078 (44)	0.24 (160)

plucked bushes were comparable with 16.9 t ha^{-1} a^{-1}. The low productivity of tea can be traced to the inefficient conversion of intercepted radiation to dry matter. The efficiency for conversion (e) can be expressed as a weight of dry matter produced per unit of radiation intercepted, and when so defined can be used to compare the performance of canopies of very different structure and leaf area index (LAI) growing in different climates. There have been no direct estimates of (e) for tea, but from the data of Othieono [21] and Magambo and Cannel [17] it appears to be about 0.6 g MJ^{-1} (PAR) for plucked tea. In comparison, the maximum measured for tropical plantation crops such as oil palm, cocoa, coconut and rubber ranges from 1.2 to 1.6 g MH^{-1}. The value for tea is therefore half that of comparable tropical perennials which themselves are about half as efficient as many temperate annual crops.

Notwithstanding this inefficiency of the canopy, the main constraint has been shown to be the very small fraction of dry matter diverted to shoots. Magambo [16] and Magambo and Cannel [17] showed that about half the dry matter was partitioned to the frame, one quarter to the extracted root system and only 10% to harvested shoots. This harvest index of 0.1 for an average yield of around 2 t ha^{-1} a^{-1} is probably the lowest recorded among the world's main crops.

Furthermore, two additional sets of data have important implications for plant selection and management:

1) Clones were shown to differ in the total amount of dry matter they produce and in the fraction of this partitioned to leaves and other organs [16, 22]. The harvest index ranged from 0.08 to 0.15 but much larger differences in productivity and partitioning must exist amongst these clones and those that have given the highest recorded yields of 8 t ha^{-1} a^{-1}.

2) Productivity was shown to be limited by the cultural system: in un-plucked compared with plucked tea, productivity (and presumably e) increased by 36%, attributed by Magambo and Cannel [17] to the larger cambial and shoot sinks that resulted from not plucking, affecting the value of (e) for this unplucked tea was around 0.9 g MJ^{-1}, only 25% smaller than the maximum recorded for crops such as cocoa and coconut.

Squire [24] showed that yield of tea is to a large extent independent of current photosynthesis and Tanton [28] concluded unequivocally that yield of tea is sink limited. An important finding on the nature of sink limitation is that the rate at which growing shoots extend is controlled mainly by the environment around the shoot itself, as defined by air temperature and potential transpiration rate (25, 29, 30), rather than by conditions in the soil. As to why so many shoots remain dormant, it is suggested that apical dominance of growing shoots may have a controlling influence through its hormonal interactions.

In commercial cultivation, the tea plant is trained to form a bush for the convenience of efficient picking of tender apical shoots at harvest, resulting in a very dense canopy. A detailed study on maintenance foliage and light interception was done by Wadasinghe [33]. His findings revealed that as much as 86% of the photosynthetically active radiation is intercepted within the first 15-cm layer of the canopy. The leaves below 25 cm from the plucking surface intercept less than 2% but comprise about 38% of the total leaf area. This clearly shows that although the tea canopy intercepts almost all the radiation received, the leaf area intercepeting the radiation is limited. Wadasinghe [33] was able to estimate the effective leaf area or the amount of leaf area exposed to photosynthetically active radiation by studying the light interception pattern. He quoted 42% as the effective leaf area of the total canopy leaf area under standard bush management practices adopted in plantations at present. Therefore, in tea, although the photosynthesis efficiency and total leaf area are comparable to other perennial crops, the leaf area capable of producing dry matter is significantly low, resulting in low dry matter production.

PREDICTION OF POTENTIAL TEA YIELD

The yield of a crop can be analysed in terms of the product of (a) solar energy received, S; (b) the fraction of this energy intercepted by the canopy, f; (c) the conversion efficiency, e; and (d) the harvest index i. S is unlikely to vary so much that it noticeably affects yield and f. will be large and uniform after the first year of the pruning cycle and will have no limitations on yield. Most of the variations between clones and fields will be in the conversion efficiency e and the harvest index i.

The consenus of physiologists is that yield increases are likely to arise through increases in the size of the sink, which may be associated with a change of either e or i or both. There is no indication as yet as to whether e or i has been responsible for the very high yields of 8 t ha^{-1} recorded in Kenya and Sri Lanka.

Squire [26] estimated the maximum potential productivity of tea taking into consideration light conversion efficiency and harvest index as follows:

On average a conversion efficiency of 0.6 g MJ^{-1} and harvest index of 0.1 give a yield of about 2 t ha^{-1}. Suppose the e cannot be increased for some reason, then the harvest index of the record crop must be 0.4, which is well within the range of maximum harvest indices of tropical plantation crops. There may be little chance of increasing it further and the potential maximum may be 8 to 10 t ha^{-1}. Alternatively, suppose that the increased sink of the record crop affected the source such that e must be 2.4 g MJ^{-1}, which is at least 1.5 times greater than the maximum efficiency measured for other tropical C_3 perennials. With the radiation available, such a crop would produce 70 t ha^{-1} total dry matter, considerably more than the record authenticated productivity, of C_3 crops in Monteith's survey. Cartwright and Cannel [2] rightly point out that the figure of 120 t ha^{-1} given by Huxley [11] is unreasonably large and suggested 60 t ha^{-1} for maximum productivity of tea, but even this is probably beyond the reach of all but very efficient tropical cereals and grasses (Table 11). Therefore, it is improbable that a yield of 8 t ha^{-1} can be obtained from a crop with a harvest index of 0.1. It could be obtained from one with a harvest index of 0.2 and a conversion efficiency of 1.2 g MJ^{-1}, and if this were the

Table 11: Estimated potential dry matter production and economic yield

Crop	Total dry matter production (t ha^{-1}a^{-1})	Yield of economic product (t ha^{-1}a^{-1})
Oil palm	44	17
Coconut	51	13
Rubber	46	15
Cocoa	56	11
Cassava	64	40
Sago	64	39
Leucaena	60	42
Tea	57 (27)	24
	120 (10)	

case, the 8 t ha^{-1} yield is probably not the potential maximum because both these values of e and i are below the maxima measured for comparable crops.

It is difficult to predict the potential maximum values of e and i for tea because its cultural system is unique. A reasonable assumption is that a tea canopy is within reach that has a conversion efficiency of 1.4 g MJH^{-1}, similar to that in the best fields has an efficiency of 0.9 g MJ^{-1} and there may be fields that give 1.0 g MJ^{-1}. If i is 0.4, the yield produced with e of 1.4 g MJ^{-1} would be about 16 t ha^{-1} year^{-1}. It is unlikeiy that tea would ever yield more than this. The realistic figure for potential maximum yield would therefore be somewhere between 8 and 16 t ha^{-1} year^{-1}.

The following suggestions have been made up crop physiologists to increase the productivity of tea:

1. Increasing the harvest index rather than increasing total biomass productivity.
   a) Older leaves could be included in the harvest as suggested by Ianton [27] or more thorough plucking with relaxation of quality constraints.
   b) Selecting clones which yield acceptable tea from leaves several phylochromes old, i.e., evaluation of clones on the basis of bud +3 leaves rather than always by + 2 leaves.
   c) Selectively harvesting on relatively shorter rounds only those shoots which are sufficiently developed.
2. Increasing shoot/root ratios. Generally high yields are obtained in low elevations because of higher shoot/root ratios. Selection of clones with greater response to fertilizer, irrigation and shade etc. might also be useful.
3. Decreasing the proportion of dry matter taken by woody stems.
   a) Plucking the proportion of dry matter apportioned to stems while not altering partitioning between leaves and roots.
   b) By keeping a very low plucking table and close spacing to maintain full canopy cover. However, this is debatable as the stems provide the subsequent crop of shoots for sustained continued production.
4. As Corley [4] suggested, breeding for less respiration rates will greatly increase the productivity of tree crops. As respiratory losses of these crops are very high, generally over 75%, even a small decrease in this loss would give a relatively large increase in net productivity.
5. Detailed studies on non-traditional yield parameters for use in breeding and selection programmes to develop high yielding clones. These parameters may include dry matter partitioning, leaf pose, canopy architecture and growth analysis parameters such as CGR, NAR, LAD (leaf longevity), RGR etc. Generic variation in photosynthesis rate has been demonstrated in oil palm, rubber and cassava. Hence, breeding for improved rates should be possible Corley [6]. Some correlations have

been established between growth of vegetative plants and photosynthesis rates. So this approach might be useful in *Leucaena* and tea where vegetative shoots are of economic importance.

6. Hadfield [7] demonstrated that the optimum temperature for maximum net photosynthesis of tea leaves is 35°C and there is no net photosynthesis when the leaf temperature is raised to 39 to 42°C. Since these temperatures are common during most parts of the year in the tropics and leaf temperatures are a few degrees above the ambient temperature, breeding of clones tolerant for higher temperatures would increase the productivity. This would also help plants to perform better during moisture stress periods.

7. Modified crop architecture by adopting different methods of plucking and bush management with a view to increase light penetration. The major portion of the canopy foliage is shaded and not effective in dry matter production. By allowing some of these leaves also to intercept light, the effective leaf area of the canopy could be improved to a greater extent.

LITERATURE CITED

1. Allen, L.H. Jr. and E.H. Lemon. 1976. Carbon dioxide exchange and turbulence in a Costa Rican tropical rainforest. In: *Vegetation and the Atmosphere* edited by J.L.Montcith New York Academic Press, pp. 265-308.
2. Cartwright, P.M. and M.G.R.Cannel. 1975, Report of a Plant Physiology Consultancy. T.R.I. of East Africa.
3. Cooper, J.P. 1975, Control of photosynthetic production in terrestrial systems. In: *Photosynthesis and Productivity in Different Environments* edited by J.P. Coopor, Cambridge University Press, pp. 593-621.
4. Corley, R.H.V. 1973, Effects of plant density on growth and yield of oil palm. *Expl. Agric.*, 9:168-180:
5. Corley, R.H.V. 1976 Photosynthesis and productivity. In: *Oil Palm Research* edited by R.H.V. Colony, J.H. Harden and B.J. Wood. Amsterdam: Elsvier, pp. 55-76.
6. Corley, R.H.V. 1983, Potential productivity of tropical perennial crops. *Expl. Agric.* 19: 217-237.
7. Hadfield, W. 1968, Leaf temperature, leaf pose and productivity of tea bush. *Nature*, 219: 282-283.
8. Hadfield, W. 1972, *two and a Bud*, 19 (2) 60-63.
9. Hadfield, W. 1974, Shade in North East India tea plantations. II. Foliar illumination and canopy characteristics. *J. Appl. Ecol.* 11: 179-199.
10. Hadfield, W. 1975. Shades in North-East Indian Tea Plantations. I, The shade pattern 22 (1) 34-60.
11. Huxley, P.A. 1975, Tea growing. *Tea in East Africa*, 15 (2):13-16.
12. Jain, J.K. 1977. Technological gaps in tea research. Technological changes in relation to productivity. *Two and a bud* 24 (2): 1-5.
13. Kira, T.,H. Ogawa, K. Joda and K. Ogino. 1967. Comparative ecological studies on 3 main types of forest vegetation in Thailand. IV. Dry matter production with special reference to the Khao Chon rainforest. *Nature and Life in South-East Asia*, 6: 149-174.
14. Kulasegaram, S.A. Kathiravetpillai and V. Shanmugarjah. 1988. Approaches to Higher

Tea Productivity through proper Bush Management. Paper presented at the Regional Tea Scientific Conference January 1988.

15. Laycock, D.H. and C.O. Othieno. Thoughts on annual dry matter production by tea. *Tea in East Africa*, **18**: 2-24.
16. Magambo, M.J.S. 1978, Crop physiology. *Tea in Eats Africa*, 18 (1): 8-10.
17. Magambo, M.J.S. and M.J.R. Cannel 1981. Dry matter production and partitioning in relation to yield of tea. *Expl. Agric.*, 17: 33-38.
18. Magambo, M.J.S. and C.O. Othieno 9177. Dry matter partitioning in tea bush. *Tea in East Africa*, **1**:15-17.
19. Monteith, J.L. 1972. Solar radiation and productivity in tropical eco systems. *J. Appl. Ecol.* **9**: 747-766.
20. Muller, D. and J. Nielson. 1965. Production brute, pertes par respiration el production nette dans la foret ombrophile tropicale. *Drt. Forstlige Forsgsvanesen Danmark*, **29**: 69-110.
21. Othieno, C.P. 1976. Annual total dry matter production in young clonal tea. *Tea in East Africa*, 16 (2): 10-12.
22. Othieno, C.O. 1982. Supplementary irrigation of young clonal tea in Kenya. 3. Comparative dry matter production and partition. *Tea*, 3(1): 15-25.
23. Sakai, S. 1975. Problem of photosynthesis of tea plant. *Japan Agricultural Research Quarterly*, 9 (2): 101-106.
24. Squire, G.R. 1977. Seasonal changes in photosynthesis of tea (*Camellia sinesis*). *J. Appl. Ecol.*, **14**: 303-316.
25. Squire, G.R. 1979. Weather physiology and seasonality of tea of Malawi. *Expl. Agric.* **15**: 321-330.
26. Squire, G.R. 1985. Ten years of tea physiology. *Tea*, 6 (2): 43-48.
27. Tanton, T.W.1977. Why tea yields are so low? *Quarterly Newsletter TRF*, **48**: 4-6.
28. Tanton, T.W. 1979. Some factors limiting yields of tea. *Expl. Agric.*, 15: 187-191.
29. Tanton, T.W. 1982 a. Environmental factors affecting yield of tea. 1. Effect of air temperature. *Expl. Agric.*, **18**: 47-52.
30. Tanton, T.W. 1982 b. Environmental factors affecting yield of tea. 2. Effect of soil temperature, daylight and dry air. *Expl. Agric.* **18**: 53-63.
31. Templeton, J.K. 1969. Where lies the yield summit for Hevea? *Planters Bulletin of Rubber Res. Ins. of Malaysia*, **104**: 220-225.
32. Visser, T. 1969. In: *Outlines of Perennial Crop Breeding in the Tropics*. Misc paper for Lanbouw Wageningen, Netherlands.
33. Wadasinghe, G. 1988. Physiology of growth and development of clonal tea (*C. sinensis* (L.)) at low elevations in Sri Lanka. Ph.D. thesis, Post Graduate Institute of Agriculture, 366 pp.
34. Weight, W. and D.W. Barua. 1975. The nature of dormancy in tea plants. *J. Exp. Bot.*, **6**: 125.
35. Witt, C.T. De, T. Brouwer and F.W. Penning De Vries. 1970. In: *Prediction and Measurement of Photosynthetic Production*. Proc. IBP/PP Tech. Pub. Wageningen PUDOC.

6

Chloroplasts Response to Low Water Potentials: Changes in Chlorophyll Fluorescence Yield and Emission of Chloroplasts as Affected by Low Leaf-water Potential

Prasanna Mohanty

Bioenergetics and Photobiochemistry Laboratory,
School of Life Sciences, Jawaharlal Nehru University,
New Delhi 110067, India

ABSTRACT

Chlorophyll a fluorescence characteristics and electron transport supported by photosystem II were monitored in sunflower and spinach leaves subjected to low moisture stress. Chloroplasts isolated from the low moisture stressed leaves did show reduced dichlorophenol indophenol Hill activity. Chlorophyll *a* fluorescence of variable yield was also reduced by low moisture stress. However, low moisture stressed leaf-chloroplasts did not show any inhibition in the ability to photoreduce exogenous electron donor like diphenyl carbazide. Low leaf water potential seem to alter energy transfer between two photosystems, as evidenced by 77 K emission spectral analysis.

INTRODUCTION

It is quite well known that low leaf-water potential inhibits photosynthesis [1]. Several partial chloroplast reactions are inhibited when the chloroplasts are isolated from leaves having low water potential [2]. These studies indicate the O_2 evolving complex (OEC) is affected by desiccation. Measurements of quantum requirements for CO_2 fixation in intact leaves and O_2 evolution with isolated chloroplasts indicate that low leaf-water potential affects the process closely linked to the primary electron transport activity of chloroplasts [3]. The O_2 evolving system is quite sensitive to a variety of external perturbations [4]. It appears that low leaf-water potential affects the oxygen evolution complex (OEC) to a much greater extent than other processes, such as electron transport photophosphorylation and CO_2 fixation.

It has already been shown that although the quantum yield of chloroplasts is reduced by low leaf-water potential, the absorption spectrum of chloroplasts does not seem to change with the low leaf-water potential in sunflower leaves [3]. Chlorophyll *a* (Chl *a*) fluorescence yield is intimately linked with electron transport and energy conversion processes of the chloroplasts [5, 8]. As Chl *a* fluorescence yield changes mostly monitor the activity of photosystem II (PSII) of photosynthesis [7, 8, 14, 16], this technique has been used to monitor PS II impairments [9].

The present article is concerned with the effect of low leaf-water potential on the Chl *a* fluorescence yield of chloroplasts isolated from desiccated leaves. The relative effect of low leaf-water potential on the chloroplasts in carrying out photosystem II (PSII) dependent photooxidation of exogenous donors as compared to photo-oxidation of water has been studied. The results indicate that although the chloroplasts lose the ability to evolve O_2 at low leaf-water potential, they none the less retain the ability to support electron flow from an exogenous electron donor via PS II. The results further indicate that low leaf-water potentials appear to alter the extent of energy transfer between two interacting photosystems.

MATERIALS AND METHODS

Leaves of three to five-week-old sunflower (*Helianthus annus*) plants or market spinach were used as the source for chloroplast preparation. Plants were desiccated in a temperature-controlled chamber as reported previously [3, 6]. Chloroplasts were isolated essentially as reported earlier [3]. Leaf-water potential was measured by a thermocouple psychrometer as previously described [3, 10] Photoreduction DCIP was measured essentially as described by Mohanty and Boyer [3].

Once-washed chloroplasts were employed for the measurements of chlorophyll (Chl) *a* fluorescence transients and emission spectra. The spectra of chloroplasts obtained from well-watered and desiccated leaf tissues were recorded in a Baush and Lomb spectrophotometer equipped with an integrating sphere to reduce errors due to scattering of the samples. The spectrofluorometer has already been described by Mohanty, *et al.* [11]. The procedure for measuring the fluorescence transients was briefly as follows: A three ml sample of chloroplasts suspension was placed in an optically flat bottom dewar flask and kept in darkness at room temperature (23° C) for 10 minutes. The dark-adapted sample was illuminated with abroad band light with a maximum transmission at 480 nm and band pass of 10 nm. The photomultiplier (EMT 9558-8 filter) was guarded from excitation source with a corning C.S. 2-64/2-58 filter. The photocurrent was amplified and recorded by the use of an oscilloscope and Easterline Angus recorder [12]. The time base scan of the scope was triggered simultaneously when the shutter of the exciting light was opened [12].

For the measurements of fluorescence emission spectra at noom and at liquid nitrogen temperature the procedures of Mohanty and Govindjee were followed [12]. The room temperature spectra were measured after the sample was illuminated for 2-3 minutes and for the liquid nitrogen spectra, the sample was maintained at 77° K during measurements. The spectra were corrected for the spectral variations of the photomultiplier and monochrometer. Other details are given in the legends to the figures such as the diffusion of CO_2 or other enzymic reaction.

RESULTS

1. Low leaf-water induced Alterations in Chlorophyll *a* (Chl *a*) fluorescence yield and emission of chloroplasts

Chlorophyll *a* (Chl *a*) fluorescence reflects the intrinsic electron transport activity of chloroplasts (7, 8). It is now well documented that changes in the yield of Chl *a* fluorescence monitors the redox state acts of the primary acceptor of PSII, the Q, which in the oxidised state acts as a quencher of Chl *a* fluorescence [5, 13]. Changes of Chl *a* emission characteristics, particularly the low temperature spectrum of chloroplasts, reflect a possible change in the spillover of energy from PSII to PSI [5, 11]. Such changes are shown to be controlled by ionic environment of the chloroplasts which regulates chloroplasts membrane structure and function (11). We, therefore, measured the changes in Chl *a* fluorescence characteristics of chloroplasts isolated from well-watered and desiccated sunflower leaves to monitor the intrinsic electron transport activity of the chloroplasts. The room temperature emission spectra of sunflower chloroplasts isolated from well-watered and desiccated (to –15 bars) leaves. The emission spectrum of chloroplasts isolated from well-watered leaves showed a typical spectrum terth main peak at 685 nm and a weak band at 725 nm. It is clear from this that intensity of the main F 685 band was suppressed by the decreases in the water potential of the leaves. The ratio of F 685/F 725 for the control sample was 6.6, while the same ratio was 3.5 for the desiccated chloroplasts. Fig. 1 shows the liquid nitrogen emission spectra of chloroplasts isolated from well-watered and desiccated (ψ–15) bars sunflower leaves. Here, again the F 685 band, which originates mostly from PSII, (12-14)

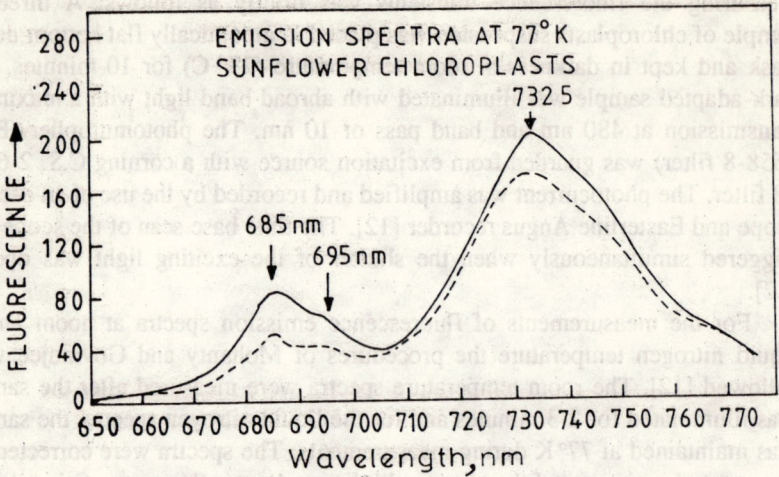

Fig. 1: Changes in Chlorophyll *a* fluorescence emission spectral characteristics at 77° K of well watered and disiccated leaf chloroplasts.

was severely suppressed. The change in ratio of short wavelength emission band (F 685) to long wavelength band in the desiccated samples might possibly be due to an increase in excitation transfer from PSII to PSI [5, 9-15].

As discussed earlier, the fluorescence induction curve is interpreted to indicate the photoreduction of electron transport carriers of the chloroplasts (5, 11, 16). It is generally assumed that increase in the relative fluorescence yield from an initial constant level (Fo), which occurs immediately at the onset of illumination, to a maximal steady state level (Fm), reflects the photoreduction of Q_A^- (16). Dichlorophenyl dimethyl urea (DCMU) or diuron, which blocks electron flow close to Q_A^- causes a rapid increase in Chl *a* fluorescence due to rapid accumulation of Q_A^- (16).

Fig. 2-A shows the fast fluorescence induction curve of chlorophyll *a* fluorescence in well-watered and desiccated leaf chloroplasts. There was only a small decrease in the initial Fo level in the desiccated sample, while the maximal Fm level was completely depressed. Fig. 2-B shows a similar kinetics when the sample was illuminated for a prolonged period. At a steady state, the (Fm-Fo)/Fo ratio was approximately 6.6 for well-watered chloroplasts and 2.4 for the chloroplasts isolated from desiccated leaves. Fig. 3 shows a similar transient curve in another chloroplast preparation where the fluorescence induction curve was measured in the presence and the absence of DCMU. Here, again, the variable fluorescence (Fm-Fo) level in the desiccated sample was low and this Fm level did not reach up to the control, Fm, level upon addition of DCMU. This suggests that the lowering of Chl *a* fluorescence by desiccation is not due to rapid cyclic reoxidation of Q_A^-, but may be due to a block in the

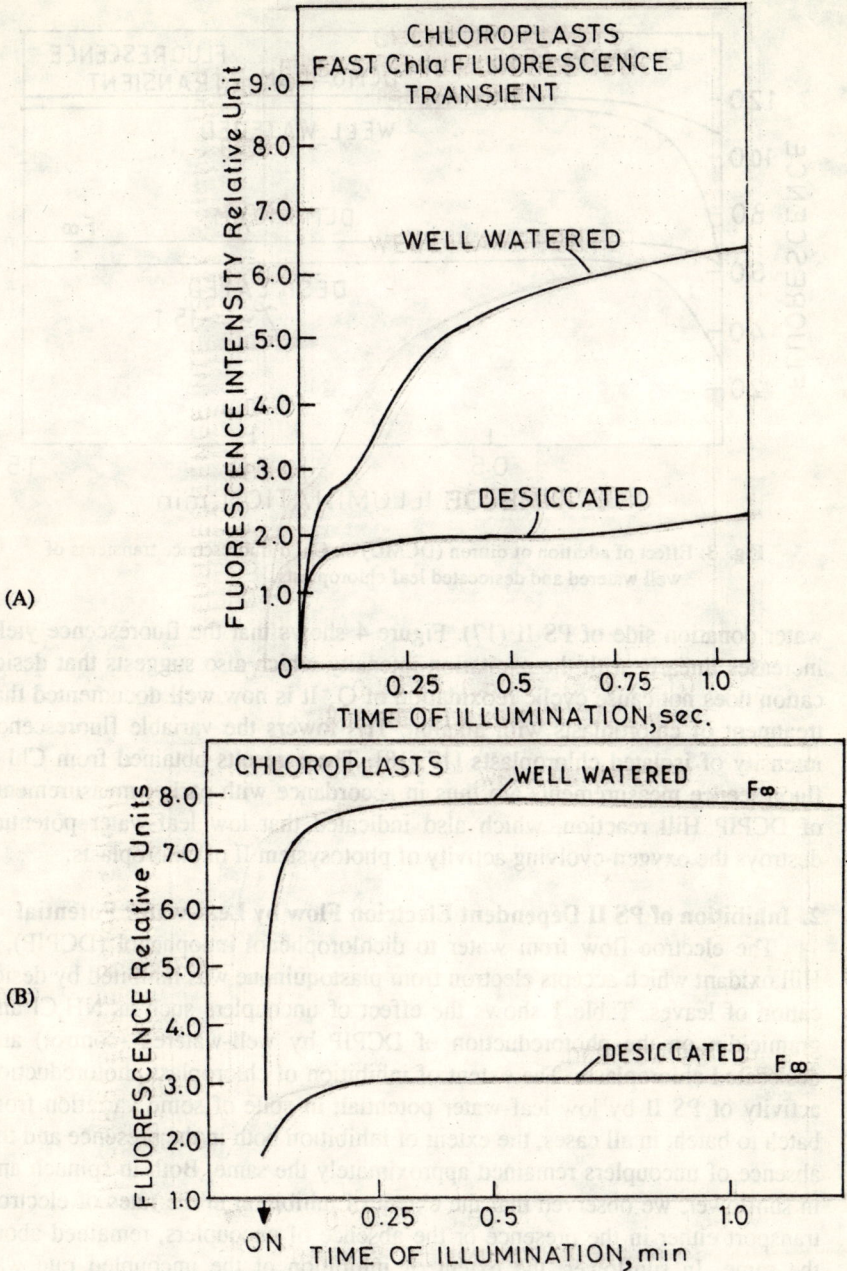

Fig. 2: Changes in Chlorophyll *a* fluorescence transients of chloroplasts by low leaf water potential.

    (A) fast fluorescence change: characteristic OIDP points are seen control chloroplasts.

    (B) Slow fluorescence transients.

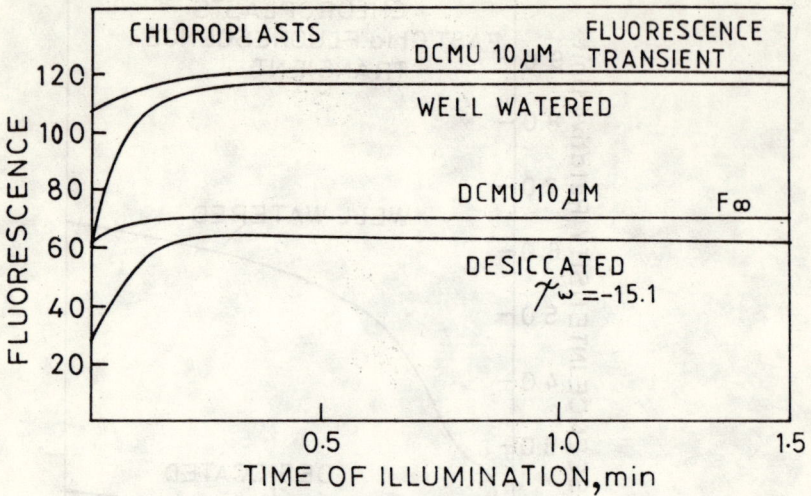

Fig. 3: Effect of addition of diuron (DCMU) on Chl *a* fluorescence transients of well watered and desiccated leaf chloroplasts.

water donation side of PS II (17). Figure 4 shows that the fluorescence yield increases linearly with the excitation intensity which also suggests that desiccation does not cause cyclic reoxidation of Q_A^- It is now well documented that treatment of chloroplasts with alkaline Tris lowers the variable fluorescence intensity of isolated chloroplasts [17, 18]. These results obtained from Chl *a* fluorescence measurements are thus in accordance with earlier measurements of DCPIP Hill reaction, which also indicated that low leaf-water potential destroys the oxygen-evolving activity of photosystem II of chloroplasts.

2. Inhibition of PS II Dependent Electrion Flow by Leaf-water Potential

The electron flow from water to dichlorophenol indophenol (DCPIP), a Hill oxidant which accepts electron from plastoquinone was inhibited by desiccation of leaves. Table 1 shows the effect of uncouplers such as NH_4Cl and gramicidin on the photoreduction of DCPIP by well-watered (control) and desiccated chloroplasts. The extent of inhibition of chloroplast photoreduction activity of PS II by low leaf-water potential, in spite of some variation from batch to batch, in all cases, the extent of inhibition both in the presence and the absence of uncouplers remained approximately the same. Both in spinach and in sunflower, we observed that the extent of inhibition in the rates of electron transport either in the presence or the absence of uncouplers, remained about the same. In sunflower, the extent of inhibition of the uncoupled rate was usually observed to be slightly more than when the uncoupler was absent.

Fig. 5 depicts the effects of the pH of the suspending medium on H_2O to DCPIP photoreduction of chloroplasts isolated from well-watered and desiccated leaves. The pH profiles for the DCPIP assay for both control and desic-

Table 1: Effect of uncouplers NH₄Cl and gramicidin on the photoreduction of DCPIP by chloroplasts isolated from control (well-watered) and desiccated leaves of spinach and sunflower

	Rate, μmol/mg Chl/hr.			% of control		
	$-NH_4Cl$	Normalised	$+NH_4Cl$	Normalised	$+NH_4Cl$	$-NH_4CL$
1. Spinach Control	273.0	100	315.0	100		
Desiccated ($\psi_0 = -27$ bars)	153.0	61	193.0	56	56.0	61.3
2. Sunflower Control	101.0	100	158.0	100		
Desiccated ($\psi_0 = -19$ bars)	42	32	48.0	42	41.6	30.4
	$-gramicidin$		$+gramicidin$		$+gramicidin$	$-gramicidin$
3. Sunflower Control	52.0	100	80.0	100		
Desiccated $\psi = -14$ bars	26.5	41	32.6	51	50.9	40.7

Conclusion: Low water potential inhibits basal electron flow.
*gramicidin, 5 μM, NH₄Cl, 2 mM.
DCPIP photoreduction was monitored under saturation conditions as detailed in the text.

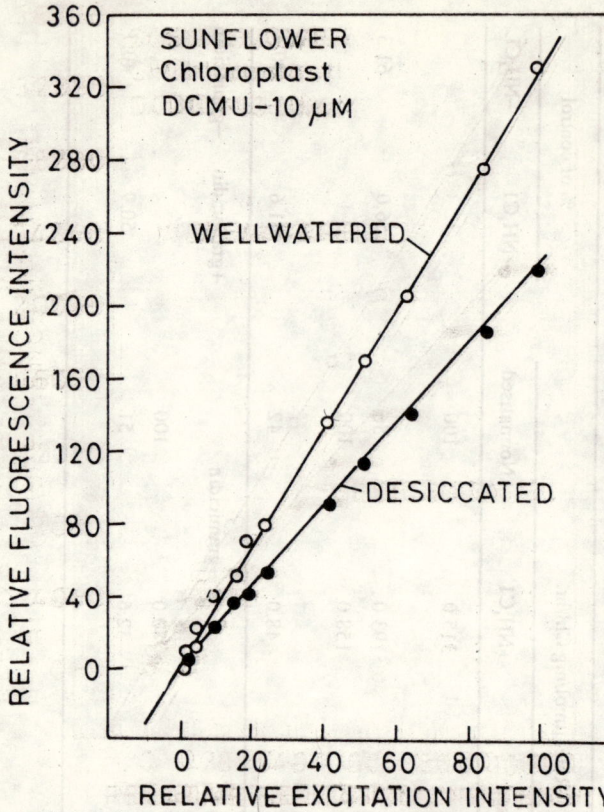

Fig. 4: The excitation intensity dependent increase in fluorescence intensity in control and draughted leaf chloroplasts in the presence of 10 μm DCMU.

cated chloroplasts remained almost the same, which suggests that the low leaf-water potential basically affects the rate of electron flow from H_2O to PS II acceptors but does not seem to alter the characteristics of electron transport carriers. There remains the possibility that the inhibition of electron transport activity of chloroplasts might be due to the formation of some as yet unidentified inhibitor in the cell cytoplasm during desiccation. Table 2 (A and B) shows the effect of incubation of spinach chloroplasts with supernatant obtained during isolation of chloroplasts from desiccated leaves of sunflower (Table 2A). The rates of DCPIP photoreduction were equally affected when the spinach chloroplasts were incubated with chloroplast supernatant of well-watered chloroplasts or with desiccated chloroplasts of well-watered chloroplasts or with desiccated chloroplasts of sunflower. Similarly, incubation of chloroplasts obtained from desiccated sunflower leaves with supernatant obtained control (well-watered) sunflower chloroplasts caused no restoration or alleviation of inhibition due to low leaf water potential (2B). It, thus, appears very unlikely that reduction of Hill activity by chloroplasts isolated from desiccated

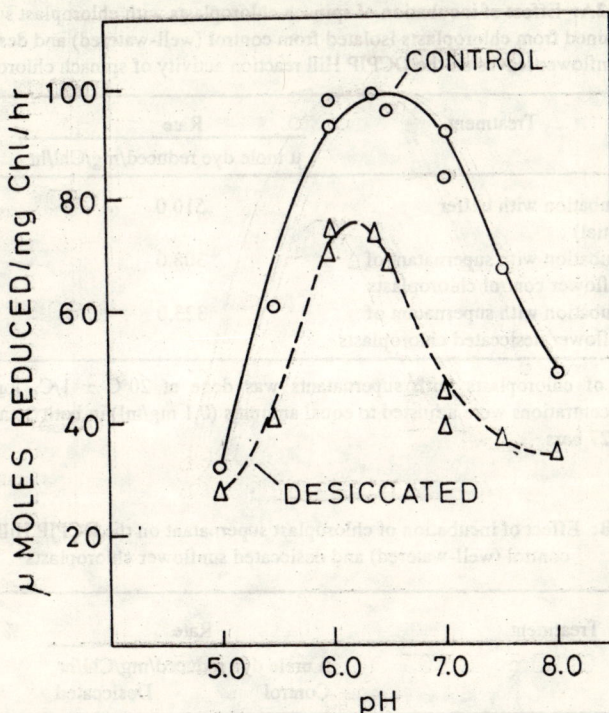

Fig. 5: pH dependent changes in Hill activity of control (well-watered) and desiccated leaf chloroplasts. pH was varied with appropriate buffers.

leaves is due to mixing of chloroplasts with any unknown cytoplasmic inhibitor during isolation.

Vernon and Shaw [19] have shown that diphenylcarbazide (DPC) is a very effective donor to PS II. They found that DPC feeds much more effectively when the O_2-evolving system is knocked out either by heating or by washing with a high concentration of alkaline Tris (18). Table 3 shows the effect of the addition of DPC (0.5 mM) on the rate of DCPIP photoreduction by control and desiccated chloroplasts. The results shown in Table 3 clearly indicate that DPC supported photoreduction in the desiccated chloroplasts showed no inhibition. Also, the addition of $MnCl_2$ to the reaction mixture caused neither an increase in DCPIP photoreduction nor a decrease in the extent of inhibition by low leaf-water potential (data not shown).

As the extent of inhibition of photoreduction activity by chloroplasts isolated from low-water potential leaves was partial, the chloroplasts were given heat treatment to inactivate O_2 evolving system (20) and were assayed for DPC-supported DCPIP-reduction. Fig. 6 shows that the heat treatment (for 5 min) given to chloroplasts inhibited electron flow from H_2O to DCPIP for both control and desiccated chloroplasts when the incubation temperature was

Table 2A: Effect of incubation of spinach chloroplasts with chloroplast supernatant obtained from chloroplasts isolated from control (well-watered) and desiccated sunflower leaves on the DCPIP Hill reaction activity of spinach chloroplast.

Serial No	Treatment	Rate	% of control
		μ mole dye reduced/mg/Chl/hr	
1.	Incubation with buffer (initial)	510.0	100
2.	Incubation with supernatant of sunflower control chloroplasts	308.0	60.3
3.	Incubation with supernatant of sunflower desiccated chloroplasts	323.0	63.3

Incubation of chloroplasts with supernatants was done at 20°C ± 1/C, for 10 min; the protein concentrations were adjusted to equal amounts (0.1 mg/ml) in both 2 and 3 desiccated leaf $\psi\psi$ = -27 bars.

Table 2B: Effect of incubation of chloroplast supernatant on the DCPIP Hill reaction of control (well-watered) and desiccated sunflower chloroplasts*

Serial No	Treatment	Rate		% of respective control
		μ mole dye reduced/mg/Chl/hr		
		Control	Desiccated	
1.	No incubation with super-natant (initial)	96.0	68.0	70.8
2.	Incubation with isolation mixture	65.3	51.5	78.8
3.	Incubation with desiccated chloroplasts superntant	68.4	48.4	70.9

*The protocol for this experiment was similar to that of Table 2A except that the activity of control sunflower chloroplasts was low. Desiccated leaf $\psi\psi$ = -19.2 bars

around 40 to 45°C; most of the photoreduction activity was inhibited by heat treatment. The addition of DPC increased the rate of DCPIP photoreduction both in control and in desiccated leaf chloroplasts. Table 4A clearly shows that after heat inactivation both well-watered and desiccated leaf chloroplasts gave equal rates of photoelectron transport activity. Table 4B supplements this observation with incubation of 0.8 M Tris-Cl (pH 8.0), a treatment which is known to affect the O_2 evolution of chloroplasts (18). After Tris-washing, both the well-watered and desiccated leaf chloroplasts yielded a high rate of photoreduction and very small or no difference in the rate of electron transport between chloroplasts isolated from control and desiccated sunflower leaves (see Fig. 6) was seen.

Table 3: Effect of diphenylcarbazide (DPC) on the photoreduction of DCPIP by chloroplasts isolated from control (well-watered) and desiccated sunflower leaves

	Rate	
	μ mole dye reduced/mg/Chl/hr	
	-DPC	+DPC
Experiment 1 Control	136.5	168.0
Desiccated		
(ψ W = -14.7 bars)	97.8	203.0
Experiment 2 Control	99.3	218.0
Desiccated	64.3	198.3
(ψ W = -20 bars)		
Experiment 3 Control	104.4	109.4
Desiccated	41.0	106.5
(ψ W = -19.7 bars)		

DPC, 0.5.mM; DCPIP photoreduction was monitored spectrophotometrically.

DISCUSSIONS

It has already been shown that chloroplast partial reactions mediated by PS II and PS I are reduced by low leaf-water potential (2, 3, 6). It was likewise shown earlier that PS II activity is more sensitive to low leaf-water potential than PS I (6). The quantum yield of H_2O to DCPIP electron transport reduced at low leaf-water potential, but the light absorption characteristics of the chloroplasts did not alter.

Both variable Chl *a* fluorescence and DCPIP photoreduction reflects the functional ability of PS II (5). The reduction of variable (Fm level) Chl *a* fluorescence by desiccated chloroplasts is reminiscent of similar results obtained

Table 4A: Effect of heat treatment on the DPC-supported photoreduction on DCPIP by the chloroplasts from control (Well-Watered) and desiccated (ψ W = − 17.0 bars) sunflower leaves

		Rate	
Heat	Treatment DPC	μ mole dye reduced/mg Chl/hr	
		Control	Desiccated (ψ = -17.0 bars)
−	−	56.7	12.6
−	+	63.3	64.8
+	−	0	0
+	+	97.2	94.0

DPC-supported photoreduction was assayed spectrophotometrically according to the procedure given in (2B)

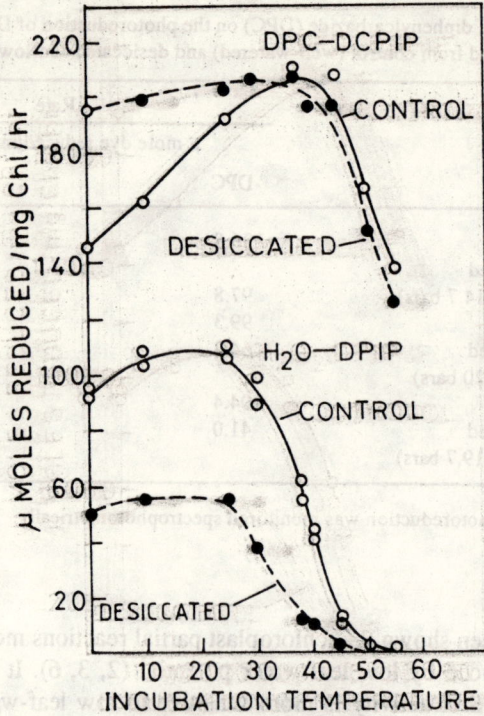

Fig. 6: Effect of temperature pretreatment on chloroplast Hill activity in the presence and absence of exogenous donor dephenylcarbazide (DPC).

Table 4B: Effect of Tris-washing on the DPC-supported Hill reaction by control (well-watered) and desiccated (ψ W = -13.0 bars) sunflower chloroplasts

Washing	Treatment DPC	Rate μ mole dye reduced/mg/Chl/hr.	
		Control	Desiccated (W = -13.0 bars)
Buffer	-	56.0	16
Buffer	+	187	161
Tris	-	0	0
Tris	+	219	167

Tris washing was done at pH 8.5 according to (ref 18); DPC, 0.5 mM.

when PS II O_2-evolving activity is damaged by heat treatment (17,20) or by alkaline Tris (17). However, unlike the case in Tris treatment, the yield of Chl *a* fluorescence did not increase significantly by the addition of 10 μM DCMU to chloroplasts isolated from desiccated leaves, which suggests that there is no endogenous source of supply of electrons to PS II centres once low leaf-water potential damages the O_2-evolving system.

Besides the changes in Chl *a* fluorescence yield, the Chl *a* emission spectrum was altered by low leaf-water potential, as evidenced by the change in ratio of emission band F 685, which primarily originated from PS II (6, 15), by F 725, the emission band associated with PSI. The emission spectrum of desiccated chloroplasts also showed that the ratio of short wavelength emission to long wavelength emission (F 685/F 730) was much lower than control (see Fig. 1). Approximately similar results were obtained when chloroplast samples were first illuminated in the presence of DCMU before freezing. The relative increase in ratio of long wavelength emission band to short wavelength band cannot be due to suppression of Q-linked variable fluorescence as evidenced from Chl *a* fluorescence transient measurements (see Fig. 3). Tris-washing of chloroplasts reduced the variable fluorescence but did not alter the Chl *a* emission characteristics. It thus appears that changes in the Chl *a* emission spectrum of chloroplasts induced by low leaf-water potential may be due to an alteration in the mode of excitation energy from PS II to PS I (14, 21).

The results on the PS II catalysed electron transport confirm earlier results that the O_2-evolving system is damaged by low leaf-water potential (3, 10).

Keck and Boyer (10) and Younis *et al.* [22] have shown that the photophosphorylation potential is maximally affected by low leaf-water potential. Our results indicate that the extent of inhibition remains the same whether the electron flow is coupled or uncoupled by the addition of an uncoupler (Table 1). It is thus clear that electron flow from water limits the photosynthetic processes and reduce the quantum yield of CO_2 fixation [3]. The possibility besides stomatal closure that desiccated leaves accumulate some sort of as yet, unidentified inhibitor which interrupts electron flow from water to the PS II reaction centre would not seem to be true (Table 2) as incubation of desiccated supernatant did not affect control activity. Although the ability of the O_2 evolving complex is apparently damaged by low leaf-water potential (Fig. 6), the chloroplasts retained their ability to photoxidise an exogenously added electron donor, such as DPC, which feeds electrons to the PS II (Table 3). Thus it appears that low leaf-water potential affects in the O_2-evolving system and reduces water oxidation capacity but not primary photochemistry of PS II.

Acknowledgement

We are grateful to Prof. J.S. Boyer, University of Illinois, for his support and suggestions. We also thank Prof. Govindjee, Plant Biology, UIUC, USA for his suggestions. Supported by grant No. FG-IN-679, IN-ARS-402 to PM.

LITERATURE CITED

1. Boyer J.S. (1976) Water deficits and Photosynthesis. In: Water Deficit and Plant Growth (T.T. Kozlowski, (editor) Vol 4, Academic Press N.Y. pp 153-190.
2. Boyer J.S. and Bowen B.L. (1969) Inhibition of oxygen evolution of chloroplasts isolated from leaves at low water potentials. Plant Physiol. 45, 612-615.
3. Mohanty P. and Boyer J.S. (1976) Chloroplasts response to low leaf water potentials IV. Quantum yield is reduced. Plant Physiol. 57, 704-079.
4. Cheniae, G.M. (1969): O_2 Evolution in Current Topics of Bioenergetics (Rao Sandi DR, Editor) Academic Press pp. 1-47.
5. Papageorgiou, G.C. (1975) Chlorophyll Fluorescence: An intrinsic probe of photosystem II. In: Bioenergetics of Photosynthesis (Govindjee, editor) Academic Press. pp 322-366.
6. Butler W.L. (1977) Chlorophyll fluorescenee as a probe for electron transfer and energy transfer. In: Encyclopedia of Plant Physiol. (A Trebst and M. Avron, editors) Vol 5, pp. 149-67.
7. Krause G.H., Lassch H. and Weis E. (1988) Regulation of thermal dissipation absorbed energy in chloroplasts as indicated by fluorescene quenching. Plant Physiol and Biochim 26, 445-452.
8. Schreiber U., Schilwa S. and Bilger U. (1986) Continuous monitoring of photochemical and non-photochemical chlorophyll fluorescene quenching with a new type of modulated fluorometer. Photosynth. Res. 10, 51-62.
9. Mohanty N. (1990) Investigation on heat induced changes in photochemistry of wheat plants (*Triticum aestium* var Kalyansana) Ph.D. thesis, JNU New Delhi.
10. Kack R. and Boyer J.S. Chloroplast response to low leaf water potential 111 differing inhibition of electron transport and photophosphorylations. Plant Physiol. 53, 474-479.
11. Mohanty P. and Braun B.Z. and Govindjee (1973) Light induced slow changes in chlorophyll *a* fluorescencee in isolated chloroplasts. Effect of magnesium and phenazine metho sultiate. Biochim. Biophys. Acta 305, 95-104.
12. Mohanty P. and Govindjee (1973) Effect of phenazine metho sulfate and uncouplers on light induced chlorophyll *a* fluorescence in intact algal cells. Photosynthetica 7, 140-160.
13. Mohanty P. and Govindjee (1973) Light induced changes in fluorescence yield of chlorophyll *a* in *Anacystis nidulans*: I Relationship slow changes with structural changes. Biochim. Biophys. Acta 305, 95-104.
14. Murata N. (1969) Cation control of energy transfer between two photosystems. Biochim. Biophys. Acta. 298, 1-17.
15. Allen J.F., Benett J. Steinback K.E. and Arntzen C.J. (1981) Chloroplast protein phorphorylation coupled PQ redox state to distribution of excitation energy between photosystems. Nature. 291, 25-29.
16. Duysens N.M. and Sweers H.E. (1963). Mechanisms of photochemical reactions in algal as studied by assays of fluorescence. In: studies on microlalque and photosynthetic bacteria. (Miyach, S. editor) pp. 353-372, Univ. Tokyo Press Tokyo.
17. Yamashita T. and Butler W. (1969) Photooxidation by photosystem II of tris washed chloroplasts. Plant Physiol. 44, 1342-1346.
18. Mohanty P, Braun B.Z. and Govindjee (1972) Fluorescence and delayed light emission in tris washed chloroplasts. FEBS Letters 20, 81-91.
19. Vernon L. and Shaw E. Photoreduction of 2-6 dichlorphenol indophenol by diphenyl carbazide. A photosystem II reaction catalyzed by tris washed chloroplasts and sub chloroplast fragments. Plant Physiol 44, 1645-1649.

20. Sabat S.C. and Mohanty P. (1989) Characterization of heat stress induced stimulation of photosystem I electron transport activity in Amaranthus chloroplasts. Effect of cations. J. Plant Physiol. 133, 686-691.

21. Origin E. (1990) Elucidation of chlorophylls fluorescence as a probe for draught stress in Willow leaves. Plant Physiol. 1280-1285.

22. Younis H.M., Boyer J.S. and Govindjee (1979) Conformation and activity of chloroplasts coupling factor exposed to chemical potential of water in cells. Biochim. Biophys. Acta 548, 328-340.

7

Effect of GA₃ on Nitrogen Fixation in some Forage Legumes of Nepal

B.N. Prasad * *and S.N. Mathur*

Botany Department
University of Gorakhpur, Gorakhpur, India

ABSTRACT

The effect of GA₃ on nodulation and nitrogen fixation in soybean (*Glycine max*), broad-bean (*Vicia faba*) and pea (*Pisum sativum*) was studied. The nodulation, nitrogen content and nitrogenase (N_2-ase) activity increased up to the 28th-day-old plant, thereafter decreased for more than a week and then increased sharply in the later growth period. GA₃ (100 ppm) increased nodulation and N_2-ase activity about 25%. In the case of pea nodulation, nitrogen content and N_2-ase activity increased gradually up to the 45th-day-old plant and decreased thereafter. Nodulation and N_2-ase activity increased 25% with 100 ppm GA₃. Soybean showed more nitrogen-fixing efficiency.

INTRODUCTION

A good legume crop can fix up to 90.8 kg. of nitrogen in a season's growth. This is equivalent to the application of 1 ton of NPK fertilizers [1]. Leguminous crops have been grown for centuries as forage grains, cattle feed and green manure in Nepal and can be used with non-legumes. But they are insufficiently used as a means of covering molecular nitrogen. The amount of nitro-

---

* Present address: Central Department of Botany, Tribhuvan University, Kirtipur, Kathmandu, Nepal.

gen fixed annually by the American soybean crop is probably more than 5 million metric tons, and by clovers in Australia more than twice this amount [2]. Chemical fertilisers are costly and furthermore, their transportation to higher elevations is difficult. Thus, soybean, broad-bean and pea have been selected for the present investigation.

MATERIALS AND METHODS

Seeds of soybean, pea and broad-bean were obtained from the Seed Corporation of India, I.A.R.I., New Delhi, and chemicals were obtained from Sigma Chemical Company and the Botany Department, University of Gorakhpur, Gorakhpur, India. Healthy seeds were selected and cleaned many times, then inoculated with specific Rhizobium species and sown in earthenware pots locally known as gamalas (30/20 cm) containing garden soil and sand 1:1 ratio. Sufficient sets of pots were made for the investigation. Fifteen days after germination, 100 ppm solution of GA_3 was sprayed on the leaves. In control sets an equal amount of water was sprayed.

Nodulation

Plants were harvested from control and treated sets from the 1st to the 33rd day after spraying and nodule number and dry weight of nodules recorded on day 0, 2, 4, 6, 8, 13, 18, 23, 28 and 33 after spraying. The data expressed is the mean of 10 plants.

Nitrogen content

Total nitrogen content of the nodules of treated and control plants was determined on day 0, 2, 4, 6, 8, 13, 18, 23, 28 and 33 after spraying, using the method of Doneen [3] and colour intensity was read with Spectronic-20 at 440 nm. The value is expressed as mg/N/g dry wt.

Nitrogenase (N_2-ase) activity

The in vivo N_2-ase activity was measured in detached nodules of treated and control plants on day 0, 2, 4, 6, 8, 13, 18, 23, 28, and 33 after spraying, using the method of Srivastava and co-authors [4]. The enzyme activity was measured using Spectronic-20 at 440 nm and expressed in terms of n mol $NH_3 10^3 h^{-1} g^{-1}$ fr. wt.

All the experiments were repeated thrice in triplicate sample set.

RESULTS AND DISCUSSION

Nodulation

Nodule numbers were maximum on the 13th day after spraying in soybean and broad-bean but in the case of pea was found on the 33rd day after spraying

(Table 1). 100 ppm GA_3 increased about 25% nodule number on the 6th day after spraying (Table 1). The dry weight of nodules increased gradually up to the 33rd day after spraying in all cases (Table 2).

Nitrogen content

Total nitrogen content in the root nodules was maximum on the 8th day after spraying in all cases (Table 3).

Nitrogenase activity

The *in vivo* N_2-ase activity measured in the root nodules ,was maximum on the 8th day after spraying in all cases (Table 4). The N_2-ase activity was increased about 25% with 100 ppm GA_3 on the 6th day after spraying (Table 4).

Table 1: Effect of 100 ppm GA_3 on nodule number of some legumes at different ages after foliar spray of 15-day-old plants. The values are mean ± SD of 10 plants

Days after spraying	Nodule number					
	Soybean		Broad-bean		Pea	
	Cont.	Treat.	Cont.	Treat.	Cont.	Treat.
0	26.1 2.70 ±	26.3 2.72 ±	30.3 2.90 ±	30.2 2.92 ±	25.0 2.60 ±	25.2 2.62 ±
2	30.0 2.90 ±	33.0 3.00 ±	32.0 3.98 ±	35.2 3.98 ±	30.0 3.70 ±	33.0 3.60 ±
4	38.2 3.98 ±	43.4 4.00 ±	38.0 3.98 ±	43.5 3.98 ±	36.0 3.70 ±	41.6 4.00 ±
6	60.1 4.65 ±	75.2 4.90 ±	62.1 4.70 ±	77.6 4.95 ±	40.0 4.00 ±	50.0 4.50 ±
8	66.0 4.70 ±	79.2 4.98 ±	67.0 4.80 ±	80.0 5.00 ±	45.0 4.20 ±	54.0 4.60 ±
13	72.1 4.80 ±	84.3 5.10 ±	76.2 4.90 ±	86.7 5.10 ±	52.0 4.58 ±	60.6 4.70
18	60.0 4.60 ±	70.2 4.80 ±	70.0 4.78 ±	81.6 5.10 ±	56.0 4.52 ±	65.3 4.65 ±
23	54.0 4.58 ±	63.0 4.60 ±	62.1 4.60 ±	72.4 4.60 ±	60.0 4.60 ±	68.5 4.80 ±
28	62.1 4.61 ±	70.3 4.75 ±	66.0 4.65 ±	75.4 4.90 ±	64.0 4.62 ±	73.1 4.90 ±
33	66.0 4.62 ±	75.3 4.85 ±	70.0 4.70 ±	80.0 5.00 ±	66.0 4.65 ±	75.4 4.85 ±

Cont. = Control, Treat. = Treated

In the present investigation, GA$_3$ (100 ppm) stimulated nodulation and *in vivo* nitrogenase activity significantly. Nandwal and Bharti [5] obtained similar results with kinetin and IAA in pea. Thus it is concluded that nitrogen-fixing efficiency can be increased significantly with 100 ppm GA$_3$ in leguminous plants

Table 2: Effect of 100 ppm GA$_3$ on dry weight of nodules of some legumes at different ages after foliar spray of 15-day-old plants

| Days after spraying | Dry weight of nodule | | | | | |
| | Soybean | | Broad-bean | | Pea | |
	Cont.	Treat.	Cont.	Treat.	Cont.	Treat.
0	24.2	24.3	31.0	31.1	28.0	28.1
2	26.8	29.0	38.2	42.0	35.0	38.8
4	34.6	40.1	48.2	55.0	42.2	48.7
6	46.8	58.5	56.0	70.2	50.0	62.5
8	52.8	59.8	64.0	72.5	58.2	67.3
13	59.6	57.9	70.2	81.6	64.4	73.6
18	62.2	69.3	74.0	84.4	68.5	77.0
23	63.6	71.6	76.2	86.7	74.2	82.4
28	65.2	73.3	80.0	90.2	76.3	85.0
33	66.4	74.8	80.0	91.2	82.2	91.3

Cont. = Control, Treat. = Treated

Table 3: Effect of 100 ppm GA$_3$ legumes at different ages after foliar spray of 15-day-old plants

| Days after spraying | Total nitrogen content mg/N/g dry wt. | | | | | |
| | Soybean | | Broad-bean | | Pea | |
	Cont.	Treat.	Cont.	Treat.	Cont.	Treat.
0	28.6	28.8	25.0	25.2	20.0	20.1
2	30.5	31.6	26.1	16.9	21.6	22.5
4	32.4	33.1	27.5	28.2	24.4	25.6
6	33.6	34.7	32.4	33.1	28.2	29.1
8	38.4	39.0	34.5	35.2	30.0	31.1
13	34.2	34.8	34.0	34.8	28.0	28.6
18	32.1	32.7	32.0	32.7	26.2	26.4
23	31.5	32.0	31.6	32.0	25.6	26.0
28	30.8	31.5	30.5	31.0	24.2	24.6
33	30.4	31.0	28.0	28.6	23.0	23.4

Cont. = Control, Treat. = Treated

Table 4: Effect of 100 ppm GA$_3$ on nitrogenase activity in the root nodules of some legumes at different ages after foliar spray of 15-day-old plants. The values are mean ± SD of three replications

Days after spraying	Nitrogenase activity					
	Soybean		Broad-bean		Pea	
	Cont.	Treat.	Cont.	Treat.	Cont.	Treat.
0	22.0	22.2	20.2	20.2	18.0	18.1
	2.6 ±	2.61 ±	2.58 ±	2.59±	2.30 ±	2.31 ±
2	30.20	33.4	26.1	28.7	23.5	25.8
	3.15 ±	3.30 ±	2.94 ±	3.10 ±	2.80 ±	2.90 ±
4	36.0	41.2	33.0	37.9	28.0	32.2
	3.48 ±	3.54 ±	3.28 ±	3.58 ±	3.00 ±	3.28 ±
6	38.5	48.1	36.2	45.3	32.4	40.5
	3.57 ±	3.70 ±	3.50 ±	3.62 ±	3.50 ±	3.58 ±
8	40.2	50.0	38.00 ±	47.2 ±	33.5 ±	41.8 ±
	3.65 ±	3.80 ±	3.58 ±	3.67 ±	3.54 ±	3.62 ±
13	38.4	47.5	36.00	43.8	31.8	38.6
	3.57 ±	3.68 ±	3.50 ±	3.60 ±	3.48 ±	3.56 ±
18	36.2	40.0	33.2	37.00	28.4	31.6
	3.48 ±	3.50 ±	3.29 ±	3.57 ±	3.05 ±	3.27 ±
23	32.00	34.8	27.50	29.00	25.00	27.2 ·
	3.15 ±	3.20 ±	2.98 ±	3.00 ±	2.90 ±	3.15 ±
28	30.00	31.4	26.00	27.2	24.00	25.2
	3.15 ±	3.22 ±	2.94 ±	3.10 ±	2.84 ±	2.94
33	24.00	24.80	21.30	21.6	20.3	21.00
	2.70 ±	2.75 ±	2.60 ±	2.65 ±	2.58 ±	2.59

Cont. = Control, Treat. = Treated

Acknowledgements

The first author (BNP) is thankful to the authorities of the Indian Embassy, Kathmandu and the University Grants Commission, New Delhi, Government of India for providing a Research Associateship in 1987. BNP is thankful to Prof. S.N. Mathur, Head, Botany Department University of Gorakhpur, for providing laboratory facilities and supervision for the present investigation. BNP is also thankful to Prof. S.C. Gupta, Department of Botany, University of Delhi, for his kind suggestions and providing laboratory facility during the course of study.

LITERATURE CITED

1.   Scott, G.D. 1969. Plant Symboisis: Studies in Biology, Edward Arnold Publishers Ltd., No. 16, 58 pp.
2.   Nutman, P.S. 1971. Perspectives in biological nitrogen fixation. *Sci. Prog. Oxf.*, **59**: pp. 55-74.

3.   Doneen, L.D. 1932. A micromethod for nitrogen estimation in plant material. *Plant Physiol.*, 7: 717-720.
4.   Srivastava, R.C., D. Mukerji and S.N. Mathur, 1980. A freeze/thaw techniques for estimation of nitrogenase activity in detached nodules of *Vigna mungo. Ann. Appl. Biol.*, 96 (2): 235-241.
5.   Nandwal, A.S. and S. Bharti. 1982. Effect of Kinetin and IAA on growth, yield and nitrogen-fixing efficieny of nodules in pea. *Indian J. Plant Physiol.*, 25 (4): 358-363.

8

Effect of *Rhizobium* Inoculation on the Nodulation and Grain Yield of Soybean in Combination with Chemical Fertilisers

Shanti Bhattarai and Surya L. Maskey

H.M.G. Division of Soil Science and Agricultural Chemistry,
Khumaltar, Kathmandu, Nepal

ABSTRACT

Rhizobium japonicum was isolated from 10 varieties of soybean. Among 10 isolates only 4 were used for the glasshouse experiment to test their response and varietal interaction on the nodulation and biomass production of soybean. The varieties selected for studies were Lumle 1, Ransom, Hill and Nuwakot Kanchhi. The *Rhizobium* inoculation increased the plant height, dried weight of nodules and plant, and number of nudules in all the varieties tested.

The most effective strain of *Rhizobium japonicum,* isolated from the Nuwakot Kanchhi variety of soybean, was used for the field response. For that a field trial was conducted at Khumaltar, Nepalgunj and Rampur to study the response of *Rhizobium* inoculation in combination with different doses of chemical fertiliser for three years. The effectiveness of *Rhizobium japonicum* was observed in non-fertilised plots; it increased the grain yield of soybean from 32–45 per cent at different locations.

INTRODUCTION

Soybean is the most potential summer legume grown in Nepal and covers about 16,000 hectares of land; yet its average yield is only 578 kg/ha [4]. Legumes, including soybean, are mostly grown on marginal land, bunds of rice terraces and in mixed cropping with maize, supplying very little or no fertiliser. Thus the legumes receive their nutrients either from soil reserves or from the atmosphere with the help of *Rhizobium*. In soil in which such bacteria are absent, using an effective strain of *Rhizobium* can increase the yield of legume.

In the present study an attempt was made to isolate *Rhizobium japonicum* from the most promising varieties of soybean. Some efficient strains were identified after efficiency tests and were used for their varietal interaction with soybean in a glasshouse. The best suitable strains thus identified were used as a bacterial fertiliser to study the response in yield and yield components of soybean in combination with different doses of nitrogen, phosphorous and potash under field conditions.

MATERIALS AND METHODS

The *Rhizobium japonicum* was isolated in a yeast extract mannitol agar (see below) from the ten most promising varieties of soybean. Among the ten isolates only four were able to form nodules on a test host in a Gibson tube having Jenson seedling agar media and these were used to study their performance on the different varieties under glasshouse conditions. Three kgs of finely ground garden soil were filled in each pot. The fertilisers nitrogen, phosphorus and potash at the rate of 20, 30, and 30 kg/ha were applied at the time of sowing seed in the form of urea, triple super phosphate and muriate of potash respectively. The treatments were replicated three times. The varieties taken for study were Lumle 1, Ransom, Hill and the local variety Nuwakot Kanchhi.

Composition of Yeast Extract Mannitol Agar (YEMA)

Mannitol	10.0 gms
Calcium Carbonate	1.0 gm
Dipotassium Hydrogen Phosphate	0.5 gm
Sodium Chloride	0.1 gm
Magnesium Sulphate	0.2 gm
Yeast Extract	0.5 gm
Agar Agar	15.0 gms
Congo Red (1:400 conc.)	5.0 ml
Distilled water	1.0 litre
pH	6.5–6.8

The media was sterilised in an autoclave for 15–20 minutes under 15 lbs pressure.

Composition of Jensen Seedling Agar (Jensen, 1942)

Calcium Hydrogen Phosphate (CaHPO$_4$)	1.0 gm
Dipotassium Hydrogen Phosphate (K$_2$HPO$_4$)	0.2 gm
Magnesium Sulphate (MgSO$_4 \cdot$ 7 H$_2$O)	0.2 gm
Sodium Chloride (NaCL)	0.2 gm

Ferric Chloride (FeCL₃)	0.1 gm
Agar Agar	15.0 gms
Distilled Water	1.0 litre
Micro-element Solution	1.0 ml

The media was sterilised as above in Gibson tubes.

After identifying *Rhizobium japonicum* as the most efficient strain from glass-house experiments, a field experiment was designed and conducted to study its effect on soybean in three different locations the terai, inner terai and hills for three years (1983-1986) in a randomised complete block design having three replications with sixteen treatments. The treatments details are as follows:

Nitrogen doses
$$\left.\begin{array}{c} 0 \\ 15 \\ 30 \\ 45 \end{array}\right\} \text{kg/ha}$$

Phosphorus
$$\left.\begin{array}{c} 0 \\ 20 \\ 40 \\ 60 \end{array}\right\} \text{kg/ha}$$

Potash
$$\left.\begin{array}{c} 0 \\ 20 \end{array}\right\} - \text{kg/ha}$$

All the three sources of fertiliser—urea, triple super phosphate and muriate of potash—were applied at the time of sowing the seed of soybean. A soil-based culture of *Rhizobium* was used at the time of seed sowing at the rate of 800 gm/ha.

Table 1: Different isolates of ***Rhizobium Japonicum*** from varieties of soybean

S. No.	Isolates of soybean	Varieties of soybean
1	66	Dare
2	68	Desotho
3	74	Dorman
4	117	Lumle local
5	127	Nuwakot Kanchhi
6	129	PI
7	94	Hill
8	205	Ransom
9	203	Kakani local
10	204	Kent

Source: Biological Nitrogen Fixation in Agriculture, 1984.

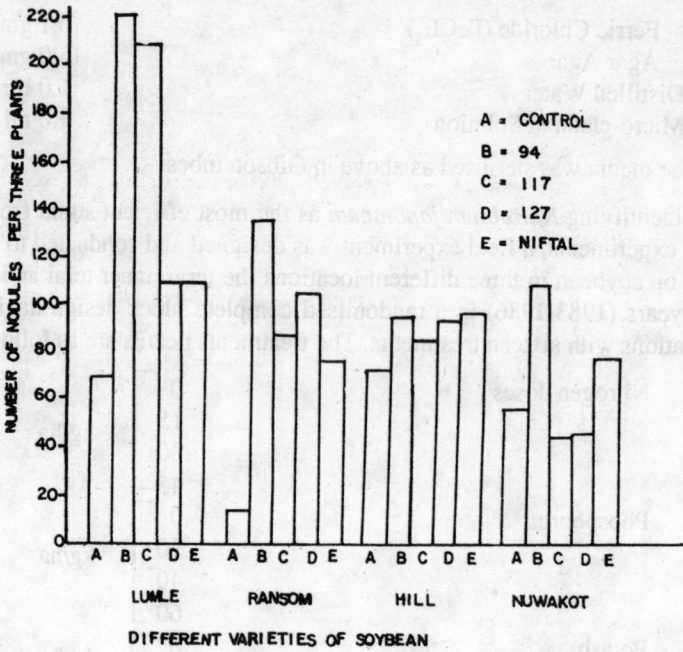

Fig. 1: Effect of different isolates on the number of nodules per 3 plants in different varieties of soybean.

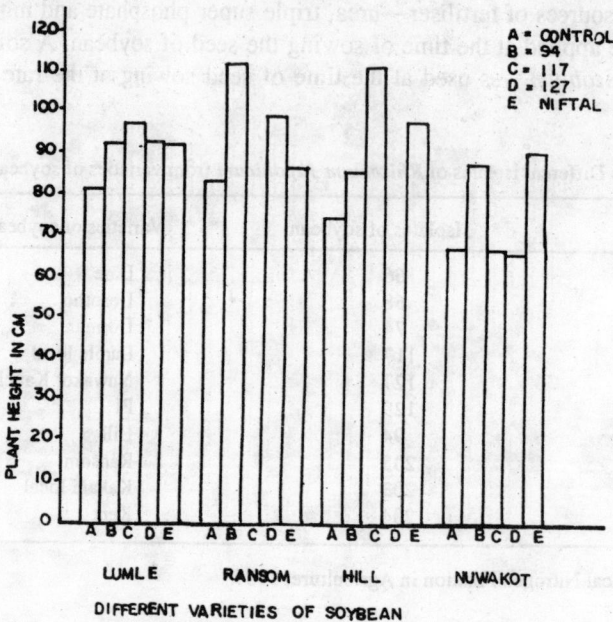

Fig. 2: Effect of different isolates on the plant height in different varieties of soybean.

RESULTS AND DISCUSSIONS

The *Rhizobium japonicum* isolated from different varieties of soybean grown in Khumaltar Agronomy farm are given in Table 1. The criteria for selecting soybean varieties for isolation were their weight, number, colour and position of nodules.

Among the 10 isolates, only four isolates, selected from four varieties, were taken to study their interaction under glasshouse conditions. The criteria for selection were their ability to form more nodules in the test plant under laboratory conditions. The results are presented in Table 2 and Figures 1-4.

Table 2. Effect of different isolates of *Rhizobium Japonicum* on the different varieties of soybean

Soybean Varieties	Treatments Isolates of *Rhizobium japonicum*	Effect on			
		Plant height (cm)	Number of Nodules/ 3 Plants	Dry Weight of plant (gm)	Weight of Nodules (gm)
Lumle	Control	81	68	29.78	0.51
	94	92	221	34.45	0.62
	117	97	209	23.82	0.66
	127	92.3	109	33.90	0.73
	Niftal	91.6	109	26.49	0.83
Ransom	Control	83.0	14	32.46	0.13
	94	111.66	135	20.76	0.54
	117	95.0	87	22.92	0.30
	127	98.33	104	24.55	0.69
	Niftal	81.0	76	21.33	0.35
Hill	Control	74.0	72	-	0.32
	94	85.0	94	22.05	0.31
	117	110.33	75	24.66	0.60
	127	88.33	93	23.33	0.65
	Niftal	92.0	96	23.30	0.75
Nuwakot Kanchhi	Control	63.66	55	30.26	0.53
	94	86.33	108	30.58	-
	117	66.60	43	13.59	0.40
	127	65.33	44	33.72	0.45
	Niftal	90.0	76	18.49	0.39

The results showed that plant height, number of nodules per plant. nodules weight and dry weight of plant increased with artificial inoculation of *Rhizobium japonicum* over inoculated control. These results accord with these obtained earlier [3]. But the behaviour of different isolates of *Rhizobium japonicum* differed with different varieties of soybean. The isolates of *Rhizobium japonicum* obtained from the hill variety of soybean gave a large number of nodules in all the varieties of soybean tested. However, the maxi-

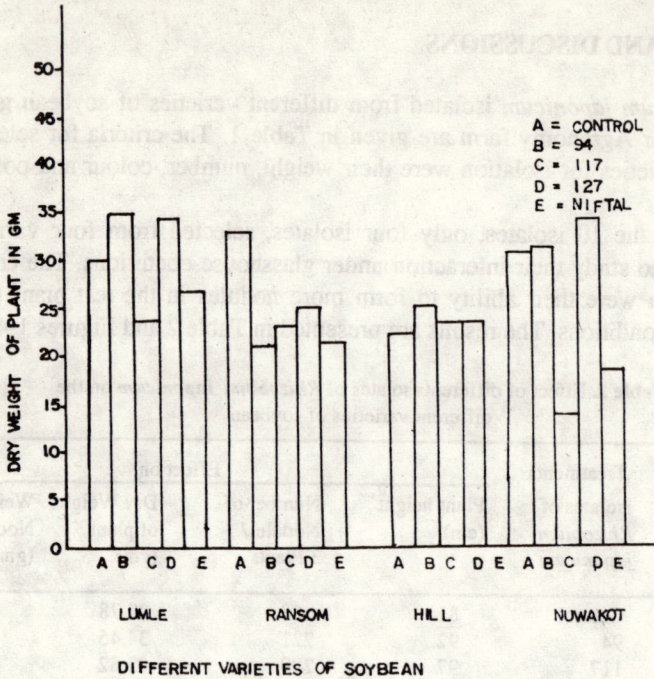

Fig. 3: Effect of different isolates on dry weight of plant in different varieties of Soybean.

mum number of nodules per plant was observed in Lumle 1 variety of soybean inoculated with *Rhizobium japonicum* isolated from the hill variety, followed by ransom, Nuwakot Kanchhi and Hill respectively. In Lumle varieties of soybean three isolates obtained from the Hill, Nuwakot Kanchhi and Nitfal responded equally in nodule formation.

The effect of *Rhizobium japonicum* in combination with and without chemical fertiliser on the yield of soybean in different agro-climatic conditions is presented in Table 3.

The grain yield of soybean obtained from all the three locations is likewise presented in Table 3. It is clear that the grain yield of soybean at Khumaltar in general was higher than at Rampur and Nepalgunj in almost all the treatments. Statistical analysis showed that the treatments were highly significant at Khumaltar, Nepalgunj but are significant at Rampur Agriculture station.

At Khumaltar no effect of nitrogen and potash was observed but the grain yield of soybean was significantly higher when 40 kg P_2O_5/ha was applied than 20 kg P_2O_5 per hectare. *Rhizobium* inoculation increased the grain yield by 45% when no fertiliser was used.

At Rampur, no effect of nitrogen, phosphorus and potash was observed in the grain yield of soybean. However inoculation of *Rhizobium* increased the grain yield by 32% over control.

Similarly, at Nepalgunj also no effect of added nitrogen, phosphorus and potash was observed in the grain yield of soybean but *Rhizobium* inoculation increased the grain yield by 43%.

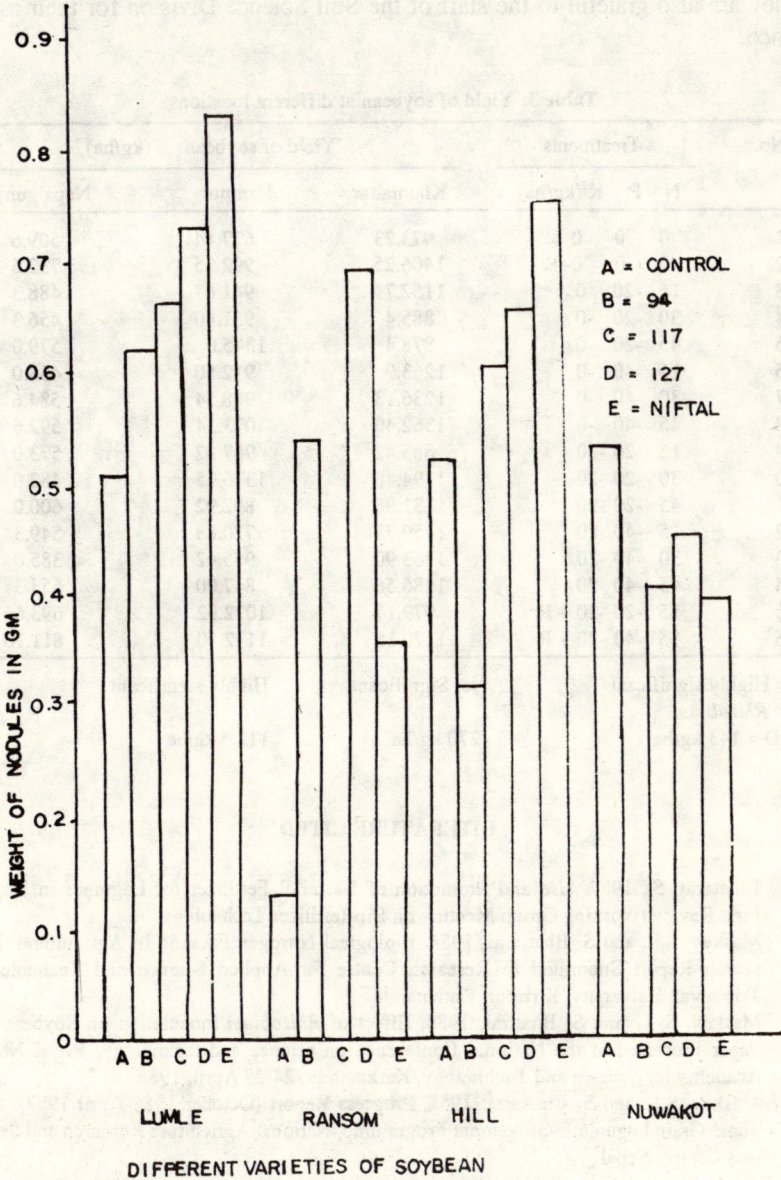

Fig. 4: Effect of different isolates on the weight of nodules in different varieties of nodules.

Acknowledgements

The authors express their sincere thanks to the Chief Soil Scientist, Mr. R. Shah, for providing facilities and valuable suggestions. They are thankful to the organisers of the Seminar for giving the opportunity to present this paper. They are also grateful to the staff of the Soil Science Division for their assistance.

Table 3: Yield of soybean at different locations

S.No.	Treatments			Yield of soybean	(kg/ha)	
	N	P	K (kg/ha)	Khumaltar	Rampur	Nepalgunj
1	0	0	0	973.23	677.44	509.6
2	0	0	0+R	1406.25	992.35	732.6
3	15	-20	-0	1152.79	941.67	488.3
4	30	-20	-0	885.4	951.40	456.3
5	45	-20	-0	878.4	1345.0	579.0
6	15	-40	-0	1263.9	912.50	438.0
7	30	-40	-0	1236.13	988.54	584.6
8	45	-40	-0	1362.40	1073.54	592.6
9	15	-20	-20	885.42	967.92	573.0
10	30	-20	-20	1194.46	1376.25	482.0
11	45	-20	-20	1131.96	892.92	600.0
12	15	-40	-20	1159.73	720.83	549.3
13	30	-40	-20	1263.90	955.42	385.0
14	45	-40	-20	1286.56	847.00	653.3
15	15	-20	-20 + R	979.17	1022.92	693.6
16	15	-40	-20 + R	1170.15	1112.50	811.16

F = Highly significant	Not Significant	Highly significant
R = *Rhizobium*		
LSD = 143 kg/ha	270 kg/ha	112.5 kg/ha

LITERATURE CITED

1.   Bhattarai, S., 1987. Use and Promotion of Bacterial Fertilizer for Legumes in Nepal. First Review/Working Group Meeting on Bio-fertilizer Technology.
2.   Maskey, S.L. and S. Bhattarai, 1984. Biological Nitrogen Fixation in Agricultural. Research Report Submitted to Research Centre for Applied Science and Technology, Tribhuvan University, Kirtipur, Kathmandu.
3.   Maskey, S. L. and S. Bhattrai, 1988. Effect of *Rhizobium* Inoculation on Soybean. A Paper Presented at the National Conference on science and technology, Royal Nepal Academy for Science and Technology, Kathmandu, 24-29 April, 1988.
4.   Maskey, S.L. and S. Bhattarai. 1988, Progress Report (October 1985-April 1987). National Grain Legume Improvement Programme, National Agriculture Research and Services Centre, Nepal.

9

Bioconversion of β-sitosterol into 17-ketosteroids by Some Strains of *Arthrobacter*

S. Mathur, M.C. Bhatia* and S.N. Mathur

Physiology and Biochemistry Research Laboratory, Department of Botany,
University of Gorakhpur - 273 009, India

ABSTRACT

Two species of *Arthrobacter*, viz., *A. simplex* and *A. globiformis* were studied to ascertain ability to degrade β-sitosterol, extracted from press mud, a sugar factory waste, into 17-ketosteroids. It was found that *A. simplex* degraded β-sitosterol into androst-4-ene-3, 17-dione (AD) and androsta-1, 4-diene-3, 17-dione (ADD) more efficiently than *A. globiformis* which needed a metabolic inhibitor, α-dipyridyl, to form AD and ADD.

Key words: Phytosterols, *Arthrobacter* Sp., Bioconversion, Androst-4-ene-3, 17-dione, androsta-1, 4-diene-3, 17-dione.

INTRODUCTION

Phytosterols have gained increasing importance as raw materials for the synthesis of steroidal drugs, besides the widely used diosgenin. Thus, all processes concerning synthesis of steroids are of considerable importance.

In the world market for pharmaceuticals, steroidal drugs represent only a fraction (1.8 billion US S of 65 billion US S). On the other hand, their various structures and pharmacological activities, as antiphlogistics, progestational—male and female sex hormones, contraceptives, and as blood-pressure regulat-

---

* Fermentation Technology Division, Central Drug Research Institute, Lucknow - 226 001, India

ing agents via their mineral-balancing influence, indicate that they are of great importance to mankind.

Diosgenin, stigmasterol and cholesterol have been used as starting materials for the synthesis of steroidal drugs, but sitosterol is a cheap and readily available natural compound. Press mud, a sugar factory waste, obtained after sulphitation of cane juice, is reported to contain 0.33% (W/W) phytosterols of which 62% is β-sitosterol, 26% stigmasterol, and 8.7% campesterol and brassicasterol [5].

A wide variety of microorganisms, such as *Mycobaceterium*, can utilise cholesterol and other phytosterols as a single carbon source for their growth and propagation [2]. *Arthrobacter* also has the property of cleaving the sterol side-chain [4, 6].

The most important step in the microbial Side-chain degradation of sterols is the break-down of the 17-ketostructure, keeping the steroid skeleton intact. Androst-4-ene-3, 17-dione was the starting material for the preparation of androgens and anabolic drugs, and as a base for the production of spironolactone. On the other hand, androsta-1, 4-diene-3, 17-dione can be aromatised by a pyrolytic procedure to estrone, which can be reduced to 19-nor-steroids. In the present study β-sitosterol was used as a starting material for the microbial conversion.

MATERIALS AND METHODS

A. Microorganisms

Arthrobacter simplex and *Arthrobacter globiformis* from the culture collection of this laboratory were used. The strains were maintained on the slants of Antibiotic Medium No. 1.

B. Cultivation Procedure

1. *Pre-culture*: Pre-culture was carried out 24 h in a temperature-controlled shaker (New Brunswick Co. Ltd.) with 120 strokes min^{-1} at 28 ± 1°C in 1000 ml Erlenmeyer flasks containing 200 ml each of two different media, viz., (i) Antibiotic Medium No. 1, and (ii) Cornsteep liquor-dextrose. The former in 1 litre contained: 6 g peptone, 4 g tryptone, 3 g yeast-extract, 1.5 g beef-extract, and 1 g dextrose. The latter in 1 litre contained: 40 g dextrose, 20 g corn steep liquor, and 1 g malt-extract. The pH of the media was maintained at 7.0 with 1.0 N NaOH.

2. *Flask Fermentation*: β-sitosterol (0.3% W/V) as substrate dissolved in ethanol and dimethyl formamide in the ratio of 3:1 were added after 6 h in four batches at an interval of 2 h each in 200 ml of the two media each containing 5% (by volume) inoculum. The inhibitor, α-dipyridyl in a concentration of 1 mM was added 8 h after initial addition of the substrate. Fermentation was carried out under aerobic conditions at 30 ± 1°C with shaker speed of 120

strokes min $^{-1}$ for 96 h.

3. *Jar Fermentation*: This was carried out in two 15 litre laboratory fermenters (Model Emenvee, Emenvee Engineers, Gultekdi, Poona, India), each containing 7 litre of the two different media with 5% (by volume) inoculum for 96 h at 30 ± 1°C with pH 7.0, aeration rate 2 litre min^{-1}, and stirrer speed at 200 rpm. The substrate (0.3% w/v) dissolved in ethanol and dimethyl-formamide in the ratio of 3:1 was added, as in the flask fermentation. The inhibitor, α,α'-dipyridyl (1 mM) was added as in the flask fermentation.

4. *Identification of the Fermentation Products*: The pure compounds isolated by column chromatography were identified by melting points, mixed melting points, Co-TLC with authentic samples, UV, IR NMR and mass spectra. UV spectra in ethanol were measured with Beckman DU-2 spectrophotometer. IR spectra in chloroform were recorded with Perkin-Elmer 2378 grating spectrophotometer. NMR spectra were taken in deuterio-chloroform solution with TMS as internal standard on Varian T 60 spectrophotometer. Mass spectra were recorded on AEA MS-30 double-focussing mass spectrometer with a Nier-Johnson geometry using an ionizing potential of 70 ev.

RESULTS AND DISCUSSION

Two different media, viz., Antibiotic Medium No. 1 and CSL-Dextrose, were used to study the conversion of β-sitosterol to AD and ADD by *A. simplex* and *A. globiformis*. CSL-Dextrose medium was found more suitable than Antibiotic Medium No. 1, as it brought about greater conversion in both species of *Arthrobacter*.

The inoculum size and the substrate concentration also affected conversion. At 0.3% substrate level, 5% (v/v) of the 24-h grown culture as seed, consumed almost the entire substrate in 96 h in the jar fermentation. On the other hand, in the flask fermentation the substrate was not entirely consumed within the same period.

The addition of 1 mM of the inhibitor, α, α'-dispyridyl after 8 h of initial addition of substrate resulted in accumulation of AD and ADD in the case of *A. globiformis*, but no accumulation of AD and ADD occurred without the inhibitor. In the case of *A. simplex*, accumulation of AD and ADD occurred even without addition of the inhibitor, but its addition increased their accumulation.

Carbon ring of the steroidal ring (ring A). The NMR signals were recorded at 0.96 (3 H, H-18), 1.26 (3H, H-19), 6.0 (H-4), 6.1d (J = 10 Hz, H-2), and 6.93d (J = 10 Hz, H-1). The same J-value for the two doublets accounted for the similar type of protons, viz., proton at C-1 and C-2 position of the steroid molecule. The IR spectra v_{max}^{CHCl3} were recorded at 2925, 2850, 1735, 1660, 1624, 1600, 1450, 1400, 1365, 1290, 1185, 1150, and 1060 cm^{-1}. The peak at 1735 accounted for a carbonyl group in the steroid (ring D) molecule. The characteristic peaks at 1660, 1624, 1600 suggested $\Delta^{1,4}$-3-C = O system of the molecule [3]. The mass peaks (m/e) were observed at 284 (M⁺, 41%), 269 (M⁺-15, 12%), 266 (32%), 251 (16%), 227 (M⁺-C₃H₄O, 23%), 199 (26%), 159 (46%), 149 (45%), 122 (100%) etc. The mass peaks at 122 accounted for cleavage of ring B, whereas peaks at 159 and 227 suggested for ring C and ring D respectively of the steroid molecule [1]. The different spectral analyses suggested the compound to be $C_{19}H_{24}O_2$. Therefore, the compound was identified as androsta-1, 4-diene-3, 17-dione (ADD), which was further confirmed by mixed melting point, CO-TLC and by complete identity of IR spectra with an authentic sample of androsta-1, 4-diene-3, 17-dione.

Acknowledgements

SM is thankful to ICAR, New Delhi for providing financial assistance and to the Central Drug Research Institute, Lucknow for providing laboratory facilities during the course of this work.

LITERATURE CITED

1. Budzikiewiez, H., C. Djerassi, D.H. Williams, and H. Day, 1964. *Structural Elucidation of Natural Products by* Mass *Spectrometry*. Acad. Press, New York, vol. II, p. 66.
2. Marsheck, W.J., S. Kraychy and R.D. Muir. 1972. Microbial degradation of sterols. *Appl. Microbial*, **23**: 72-77.
3. Nagasawa, M., M. Bae, G. Tamura, and K. Arima, 1969. Microbial transformation of sterols. Part II. Cleavage of sterol side-chain by microorganisms. *Agric. Biol. Chem.*, **33**: 1644-1650.
4. Nagasawa, M., N. Watenabe, H. Hashiba, M. Murakame, M. Bae, G. Tamura, and K. Arima, 1970. Microbial transformation of sterols. Part V. Inhibitors of microbial degradation of cholesterol. *Agric. Biol. Chem*, **34**: 838-844.
5. Srivastava, R.A.K., S.K. Srivastava, and S.N. Mathur, 1983. Isolation of β-sitosterol from sugarcane waste and its bioconversion into ADD using *Arthrobacter oxydans. Curr. Sci.*, **52**: 823-824.
6. Srivastava, S.K., R.A.K. Srivastava, and S.N. Mathur, 1985. Biotransformation of sugarcane sterol into androsta-1, 4-diene-3, 17-dione (ADD) by *Arthrobacter globiformis* Str. oxydans. *J. Appl. Bacteriol.*, **59**: 399-402.

10

Regeneration of Plants from Leaf Explant in Orchid *Vanda teres* Lind

R. Niraula and S.B. Rajbhandary

Botanical Survey and Herbarium, Dept. of Medicinal Plants,
Kathmandu, Nepal

ABSTRACT

Explants of *Vanda teres* Lind were excised from aseptically grown seedlings and cultured on Murashige and Skoog medium supplemented with auxin and cytokinin. Protocorm-like bodies were depeloped from the explants. These protocorm-like bodies were transferred to a Vacin and Went medium and grew into complete plantlets.

INTRODUCTION

Clonal propagation of orchids' through tissue culture was first attempted by Morel [1, 2], who cultured the meristem of *Cymbidium* hybrids. Now the clonal propagation of orchids by shoot tip or meristem culture is a well-established technique in such orchids as *cattleya, Cymbidium* and *Dendrobium*—a technique that has revolutionised the orchid industry. In this technique the most important growing part of the plant has to be sacrificed. Attempts to raise seedlings from various vegetative parts have been successful with only a very few orchids [3]. The present work describes the method for culturing excised leaf explants of *Vanda teres* Lind. This is a warm-climate orchid with a slender stem and terete leaves. The plant flowers during the dry period from March to April, bearing 2 to 4 large rose-coloured flowers, 10 cm across, which are long-lasting and fragrant. The orchid is also widely used to produce hybrids.

MATERIALS AND METHODS

The leaves about 10 cm long, were excised from 6–8-month-old seedlings aseptically grown in Vacin and Went medium [4]. The leaf pieces, about 5-8 mm (approx.), were cut down with a sterile blade. Two to three leaf explants were cultured on the solidified Murashige and Skoog (MS) medium [5] supplemented with indoleacetic acid (IAA), kinetin and 1000 mg/l of casein hydrolysate. The pH of the medium was adjusted to 5.8 before autoclaving.

The cultures were incubated at 25°C under 16 hours photoperiod at 3000 lux.

RESULTS AND DISCUSSION

The leaf explants cultured in MS medium supplemented with auxin and cytokinin initially remained green with no apparent changes or growth (Fig. 1). After 8 weeks of culture, the explants cultured in a higher concentration of auxin and cytokinin (5 mg/l of IAA and 5 mg/l of kinetin) thickened, especially at the cut ends, and small greenish dot-like protuberances or small papilla-like structures developed. After 3 months of culture the explants turned dark brown. The papillae developed into a green mass of protocorms attached to the explant after 4 months of culture (Fig. 2).

The protocorms when subcultured in Vacin and Went medium, differentiated into seedlings by developing the first leveas in 4-6 weeks. The first roots developed after this. Thus the complete seedlings developed 6 months after the leaf explant culture (Fig. 3). Further growth of the seedlings occurred and more roots developed. Complete plantlets were obtained after 6 to 8 months of culture, which were transferred to an epiphytic medium under greenhouse conditions (Fig. 4).

The leaf explants were cultured in a combination of different concentrations of IAA and kinetin ranging from 1 mg/l to 5 mg/l, Protocorm formation occurred only in the medium containing a higher concentration (IAA 5 mg/l and kinetin 5 mg/l).

The results indicate that leaf explants can be used for clonal propagation. Propagation of orchids from leaf explants offers several advantages. Firstly, the method is useful in those cases where seeds are not available. Secondly, the procedure is very simple and the explants are taken from the less valuable part of the plant. Furthermore, all the portions of the leaf more or less form protocorms. By maintaining the culture through several subcultures a virus may also be eliminated [6].

Given this kind of culture practice, it appears possible that other commercially important *Vanda* hybrids would show similar growth potential and might prove helpful in increasing the production of desired clonal plants.

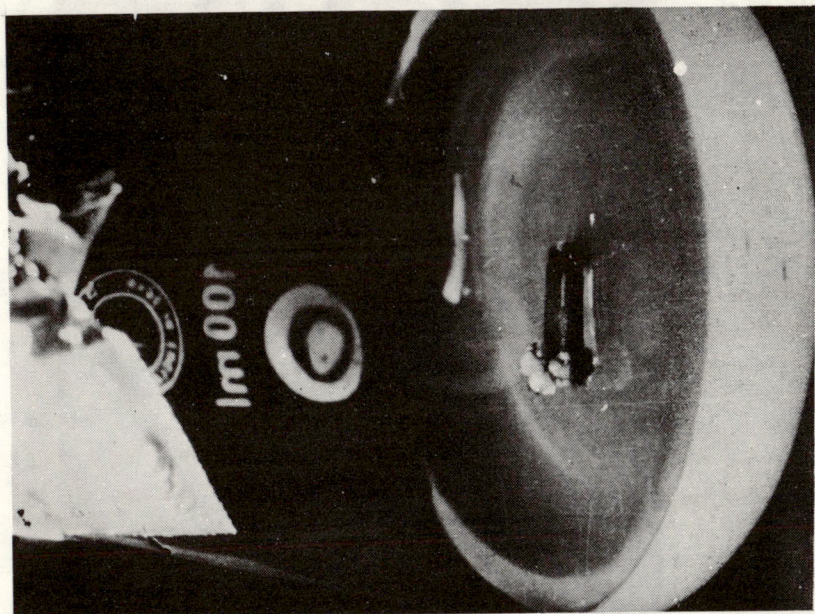

Fig. 2: Development of protocorms.

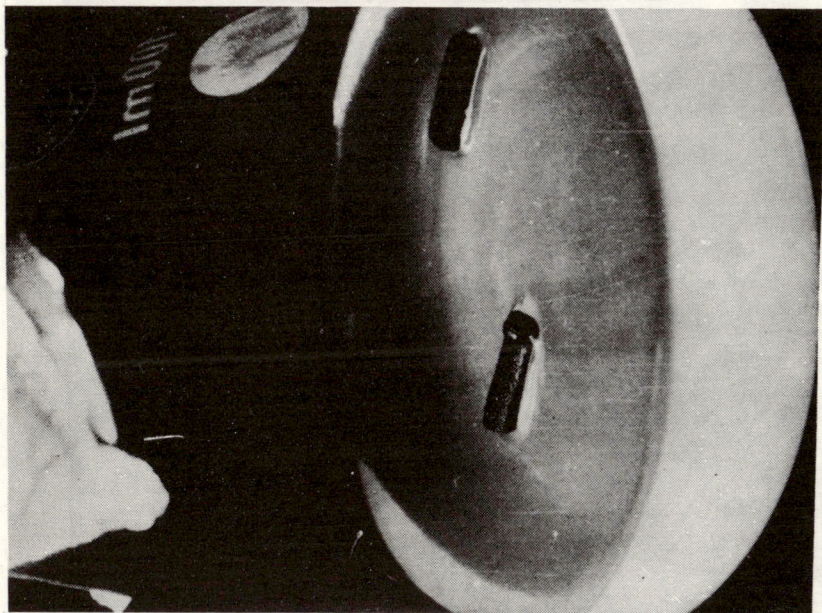

Fig. 1: The leaf explant.

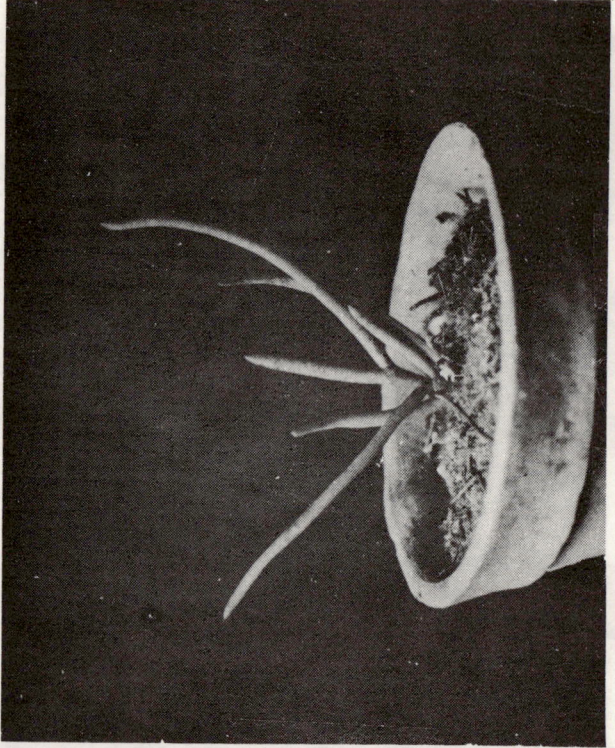

Fig. 4: Established plant.

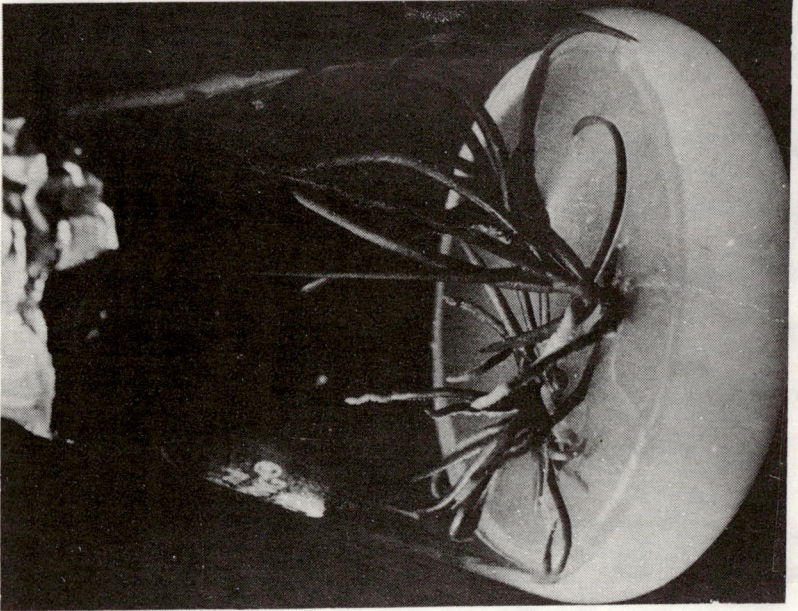

Fig. 3: Well-developed seedlings.

Acknowledgements

We are very thankful to Dr. S.B. Malla, Director General Department of Medicinal Plants, Ministry of forest & Environment, H.M.G., Nepal for providing facilities for the study.

LITERATURE CITED

1. Morel, G. 1960. Producing virus-free Cymbidiums. *Amer. Orchid Soc. Bull.,* **29**: 495-497.
2. Morel, G. 1964. Tissue culture—a new means of clonal propagation of orchids. *Amer Orchids. Soc. Bull.,* **33**:473-478.
3. Murashige, T. and P. Skoog. 1962. Revised medium for rapid growth and bioassays with tobacco tissue cultures. *Physiol. Pl.,* **15**: 473.
4. Rao, A. N. 1977. Orchid industry. *Applied and Fundamental Aspects of Plant Cell Tissue and Organ Culture.* J. Reinert and Y.P.S. Bajaj (eds.) Springer-Verlag, Berlin, pp. 44-65.
5. Vacin, E. and F. Went. 1949. Culture solution for orchid seedlings. *Bot. Gaz.,* **110**: 605-613.
6. Wang, P.J. 1977. Regeneration of virus-free plants from tissue culture. In: *Plant Tissue Culture and* Its *Biotechnological Application* edited by W. Bars, E. Reinhard and M.H. Zenk. Springer-Verlag, Berlin, pp. 386-391.

11

Generation of Genetic Variability by the Chrysanthemum Tissue Culture Method

Asha Karki

Department of Forestry and Plant Research, Royal Botanical Garden
Tissue Culture Laboratory, Godawary, Kathmandu, Nepal

ABSTRACT

Plants regenerated through tissue culture were established for various explant sources such as leaf, pedicels, flowers, buds and petals. Plants from petal explants were phenotypically different when coupled with the parental plants. Upon analysis, these regenerated plants differed in relation to the height of the plant and flower morphology. So, it was found possible to generate genetic variability in *Chrysanthemum*.

INTRODUCTION

The clonal propagation of plants through tissue culture began with an ornamental species, viz., Cymbidium [1]. The regeneration of plants has been successfully applied to a large number of families and genera [2]. Stewart and Dermen [3] were the first to report plant regeneration from shoot tips of the Chrysanthemum 'Indianapolis'. The regenerates exhibited differences in flower colour, which apparently arose from a breakdown of the chimeral nature of the apical meristem.

Sulter and Langhans [4] regenerated plants from a nine-year-old callus of *C. morifolium* of Ramat. These workers observed morphological abnormalities in these callus-derived plants compared to plants from a one-month-old callus. The implication from their results, both for research and the commercial appli-

cation of tissue culture of chrysanthemum, is that the generation of new cultivars, using tissue culture procedure, would be more effective with a culture
maintained for a nine-year-old callus.

In this paper the regenerative potential of tissue explants of *C. morifolium*
is assessed and the best explant source for plant regeneration suggested.

MATERIALS AND METHODS

i)   Fully expanded leaves from 60-65-day-old plants *C. morifolium* geno-
      types EC-7, EC-8, EC-17, EC-23 grown under greenhouse conditions
      were used as source material. They were sterilized with 8% Domestos for
      20 minutes followed by six washes with sterile tap water.

ii)   *Pedicel*: Pedicel segments of *C. morifolium* genotypes EC-7, EC-8, EC-
      17 and EC-23 from 45-50-day-old plants were taken from greenhouse
      grown plants and were surface sterilised with 10% (v/v) for 15 min, fol-
      lowed by six washes with sterile tap water.

iii) *Flower bud*: Mature buds (2.0 cm) and young buds (0.5 cm) of *C.
      morifolium* genotypes EC-7, EC-8, EC-17 and EC-23 were taken from the
      greenhouse-grown plants and surface sterilised with 10% (v/v) for 15 min,
      followed by six washes with sterile tap water.

iv)   *Petal explant*: mature petals of four genotypes of *C. morifolium* were
      detached from flowering plants grown in the greenhouse and were surface
      sterilised as for flower buds.

After suface sterilisation, explants were kept on Murashige and Skoog
medium [5] at different concentration of Kinetin and NAA. The Media was
solidified with 0.8% agar and the pH was adjusted to 5.8.

For an evaluation of morphogenetic response of explants, cultures were
maintained under continuation illumination (2000 lux) at 23 ± 2°C. After 6
weeks, the callus induction response as well as morphogenetic response of the
explant tissues were assessed for each genotype.

RESULTS AND DISCUSSION

The effect of BAP and IAA on the morphogenesis of pedicel explants of the
four genotypes of *C. morifolium* is presented in Table 1. All explants exhibited
direct shoots after two weeks. A comparison of the four genotypes seems to
reflect a genotypic influence on callus formation and plant regeneration. EC-23
produced shoots in all combination of BAP and IAA but the genotypes EC-8,
EC-7 and EC-17 did not. Only callus was produced in all cases. Maximum
proliferation in EC-23 was with BAP (5.0 mg/*l*) and kinetin (1.0 mg/*l*).

The effect of the two phytohormones, 6-BAP and IAA, on the morphogenesis from flower buds of the four genotypes of *C. morifolium* was observed
(Table 2).

Table 1: Effect of BAP and IAA on morphogenesis of pedical explants of four
C. morifolium genotypes

Genotype	Explant size	Phytohormones (mg/1)			(BAP+IAA)	Basal medium
		5.0 + 1.0	2.5 + 1.0	1.25 + 1.0	0.65 + 1.0	0.25 + 1.0
EC-7	0.5 cm	C	C	C	-	C
	1.0 cm	C	C	C	-	C
	2.0 cm	C	C	C		C
EC-8	0.5 cm	BC	BC	CS	BC	C
	1.0 cm	BC	BC	CS	C	C
	2.0 cm	BC	BC	CS	C	C
EC-17	0.5 cm	CS	CS	C	C	C
	1.0 cm	CS	C	C	C	C
	2.0 cm	CS	C	C	C	C
EC 23	0.5 cm	CS+++	CS++	CS++	CS+	CS+
	1.0 cm	CS+++	CS++	CS++	CS+	CS+
	2.0 cm	CS+++	CS++	CS++	CS+	CS+

Key: S+++   =   8-10 shoots per (0.5-2.0 cm) explant
       S ++   =   4-6 shoots per (0.5-2.0 cm) explant
       S +   =   2-6 shoots per (0.5-2 cm) explant

The effect of NAA and kinetin on the morphogensis of petal explants of four genotypes of *C. morifolium* is presented in Table 3. After 4 weeks on an MS medium supplemented with NAA (2.0 mg/*l*) and kinetin (10.0 mg/*l*) calluses appeared from the basal portion of the petal (Fig. a) and gradually lost their colour. Shoots were produced during callusing from the basal region of each petal. Production of callus was observed on the margin of the petals above the basal region. In order to sustain plant production, the regenerating petal was transferred to the same medium after four to six weeks. After another four to six weeks on average 16-18 shoots per petal were produced (Fig. b).

Table 2: Effect of BAP and IAA on morphogenesis of flower buds of four
C. morifolium genotypes

Genotype	Explant	Phytohormones (mg/l) MS basal medium				(BAP + IAA)
		0.65 + 1.0	1.25 + 1.0	2.5 +1.0	5.0 + 1.0	MSO
EC-7	Young flower buds	S+	S++	S++	S+++	C
EC-8	Young flower buds	S+	S+	S+	S++	C
EC-17	Young flower buds	S+	S+	S+	S++	C
EC-23	Young flower buds	S+	S++	S++	S+++	C

Key: S+   =   shoot regeneration with number of shoots per flower bud   15
       S++   =   shoot regeneration with number of shoots per flower bud   16
       S+++ =   shoot regeneration with number of shoots per flower bud   20

Table 3: Effect of NAA and kinetin on morphogenesis of petal of *C. morifolium*

Geno-type	Normal Flower colour	N + K MSO	N + K 2+1	N + K 2+2	N + K 2+3	N + K 2+5	N + K 2+10	Intensity of shoot formation
						(NAA+KN) N +K		
EC-7	Crimson red	-	CR	CR	C	C	CS	+
EC-8	Yellow	-	C	CR	CR	CS	CRS	++
EC-17	Yellow	-	CR	CR	CRS	CRS	CS	+++
EC-23	White	-	C	C	C	CS	CS	++

Key: +++ = 16-18 shoots per explant
      ++ = 6-8 shoots per explant
      +   = 4-6 shoots per explant
      -   = No response

Adventitious roots were initiated on MSO medium after 10-15 days of transfer from the regeneration medium (Fig. c). After four weeks the rooted plants were transferred to pots (Fig. d).

There are several investigators who have been able to regenerate plants from different cultured explant tissue of Chrysanthemum. In this plant, floral organs are more responsive in culture than other explant tissue.

Roost and Bokelmann (1975) used various concentration of BAP and IAA pedicle shoots regeneration to induce found a combination of 1.0 mg/*l* BAP and 1.0 mg/*l* IAA in MS medium suitable stimulate shoot production. In our experiments, 5.0 mg/*l* BAP and 1.0 mg/*l* IAA were found to induce the shoot proliferation.

Bush et al. (1976) obtained shoot regeneration from petal explants of *C. morifolium*. They observed abnormal morphological characters in comparison to parental plants, the regenerated shoots exhibiting the quilled and incised forms of petal as well as lack of anthocyanin pigment. The changes in morphology of plants derived from *in vitro* cultures may actually be caused by mutations occurring as a result of tissue culture (Heinz and Mee, 1971; Wright and Northcote, 1973).

It was observed that the plant regeneration capability not only differs among plants of different genotypes, but is also dependent on the nature of the source of explants of the same plants. Explants of flower buds and petals have a greater regenerative capability than other plant parts.

Acknowledgements
I am sincerely grateful to the Director General, Dr. S.B. Malla and Deputy Director General, Dr. S.B. Rajbhandari for the opportunity to conduct this study in the U.K. and to the British Council for financial support. I also thank Mr. B. Case for his expert assistance in preparing photographs.

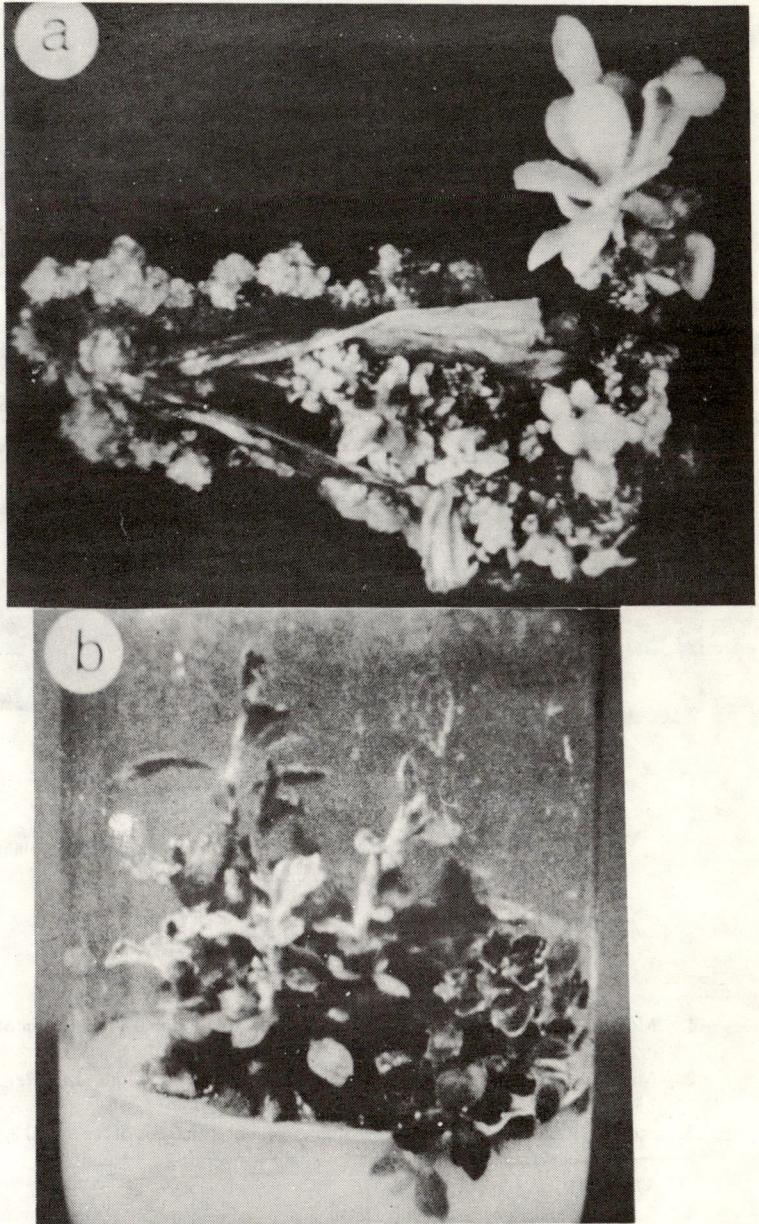

Fig. a: Shoots produced from a petal explant.

Fig. b: Profuse regeneration of shoots after subculture of a petal explant
to new medium ($\times 0.8$)

Fig. c: A regenerated shoot produced roots (× 13)

Fig. d: A regenerated plant derived from a petal explant. The plant is shown 8 weeks after transfer to putting comfort (× 0.1).

LITERATURE CITED

1. Morel, G.M. 1964. Tissue culture—a new means of clonal propagation of orchids. *Amer. orchid Soc. Bull.*, **33**: 473-478.
2. Murashige, T. 1974. Plant propagation through tissue culture. *Ann. Rev. Plant Physiol.;* **25**., 135-166.
3. Stewart, R. and H. Dermen. 1970. Somatic genetic analysis of the apical layers of chemeral spots in chrysanthemum by experimental production of adventitious shoots. *Am. J. Bot;* **57**., 1061-71.
4. Sulter, E. and R.W. Langhans. 1981. Abnormalities in chrysanthemum regenerated from long-term cultures. *Ann. Bot.*, **48**: 559-568.
5. Murashige, T. and F. Skoog. 1962. A revised medium for rapid growth and bioassays for tobacco tissue cultures. *Physiol. Plant*, **15**: 473-497.

12

In vitro Morphogenetic Studies in *Amaranthus paniculatus* L.

C.M. Govil

Botany Department, Meerut University, Meerut-250005, India

ABSTRACT

The explants of root, shoot, leaf, hypocotyl and cotyledons of *Amaranthus paniculatus* responded well to the induction of callus on different basal media [2-6] supplemented with different phytohormones. However, MS basal medium with NAA (1, 5 mg l^{-1}) + 2, 4-D (1 mg l^{-1}) + K (0.05 mgl^{-1}) was found to be most optimal for callus induction and growth. Direct differentiation from explants and from callus could then be obtained by regulating the hormonal concentrations in the MS basal medium. High auxin concentrations, IAA and IBA (5, 10 mg l^{-1}) induced roots and high concentrations of K and BAP (1, 2 mgl^{-1}) induced shoots. The shoot buds, on transfer to rooting media, produced profused rooting from the cut ends of the shoot, which could be transferred to the field. Plants *in vitro* produced profused flowering when incubated in light. The survival of such plants was 90%.

INTRODUCTION

Among the better known wonder crops of India, the grain Amaranth (*Amaranthus paniculata*) is one which has high economic value as food because of its high protein content. The crop provides scope for improvement but no serious attempts have been made in this direction to date. Plant, cell, tissue and organ culture techniques can be of great advantage in isolating somaclone variants of desired characters. Therefore, the present studies were undertaken on the various aspects of growth and differentiation *in vitro* of tissue of *A. paniculatus*, in order to standardize the minimal requirements for these processes.

MATERIALS AND METHODS

Seeds of *Amaranthus paniculatus* were procured locally. These were surface-sterilised with 0.1% mercuric chloride, rinsed 3-4 times with sterilised double distilled water and grown aseptically on 0.6% water-agar medium at 26 ± 2°C. Different explants were inoculated on different basal media supplemented with different phytohormones. The pH 5.6 was adjusted before autoclaving at 1.06 kg/cm² for 15 minutes. Tissues were incubated at 26 ± 2°C ion dark or in continuous fluorescent light, 2000 lux. The callus was transferred on MS basal medium supplemented with different concentrations of IAA, NAA, K and BAP for differentiation.

RESULTS

A. *Callus Induction and Growth*: Induction of callus was studied using different nutrient media supplemented with phytohormones (Table 1). The results showed that of all the nutrient basal media, MS medium supplemented with NAA (2, mg 1⁻¹), + 2, 4-D (2 mg 1⁻¹) + K (0.05 mg 1⁻¹) was optimal. Increased cyhtokinins enhanced callus induction and growth. Growth of callus in terms of fresh and dry matter weights was maximum after five weeks of incubation on MS medium with NAA (1 mg 1⁻¹) + 2, 4-D (5 mg 1⁻¹) + K (0.05 mg 1⁻¹) (Fig. 1).

B. *Differentiation*: Differentiation of root and shoot could be obtained from direct explant or from callus. Shoot and root differentiation from hypocotyl and cotyledon explants inoculated on MS medium with NAA or IAA (1, 2, 5 mg1⁻¹) + K (0.05 mg1⁻¹) or MS with K or BAP (1, 2 mg1⁻¹) + IAA or NAA (0.5 mg 1⁻¹) was good and more than 70% cultures showed shoot differentiation (Table 1, Fig.2). From callus also both root and shoot differentiation could be obtained. Profused rooting was observed on all the media containing basal MS + NAA (5 mg 1⁻¹) + K (0.05 mg1⁻¹) about 70-80% cultures showed root differentiation. 2, 4-D inhibited root differentiation (Table 1).

Differentiation of shoot was observed on MS basal medium supplemented with varying concentrations of auxins and cytokinins. More than 90-95% cultures showed profused plantlet formation on this medium. However, K (0.05 mg1⁻¹) in combination of 2,4-D (1, 2, 5 mg1⁻¹) completely inhibited shoot formation. High concentrations of cytokinin with low auxin content enhanced shoot formation. K (2 mg1⁻¹) + NAA (0.5 mg1⁻¹) showed the best results (Table 1).

80% of cultures showed root differentiation when the shoots were transferred on a rooting medium, NAA (1, 5 mg1⁻¹) + K (0.05 mg1⁻¹). IAA (2, 5 mg1⁻¹) + K (0.05mg1⁻¹) showed better rooting of the plantlets (Fig. 3). When

Table 1: Response of callus initiation and growth to varying auxin concentrations in Heller's, Nitsch's, White's B₅ media.

S.No.	Media supplemented with different concentrations of auxin mgl⁻¹	Cotyledon	Hypocotyl
	Heller's medium		
1.	H + 0.05 K + 1 NAA	-	+
2.	H + 0.05 K + 2 NAA	+	+
3.	H + 0.05 K + 5 NAA	+	++
4.	H + 0.05 K + 1 NAA + 1, 2, 4-D	+	+
5.	H + 0.05 K + 1 NAA + 2, 2, 4-D	+	+
6.	H + 0.05 K + 1 NAA + 5, 2, 4-D	++	++
	Nitsch's medium		
1.	N + 0.05 K + 1 NAA	+	+
2.	N + 0.05 K + 2 NAA	+	+
3.	N + 0.05 K + 5 NAA	+	+
4.	N + 0.05 K + 1 NAA + 1, 2, 4-D	+	+
5.	N + 0.05 K + 1 NAA + 2, 2, 4-D	+	+
6.	N + 0.05 K + 1 NAA + 5, 2, 4-D	+	++
	White's medium		
1.	W + 0.05 K + 1 NAA	-	-
2.	W + 0.05 K + 2 NAA	-	-
3.	W + 0.05 K + 5 NAA	+	+
4.	W + 0.05 K + 1 NAA + 1, 2, 4-D	+	+
5.	W + 0.05 K + 1 NAA + 2, 2, 4-D	+	+
6.	W + 0.05 K + 1 NAA + 5, 2, 4-D	+	+
	B₅ medium		
1.	B₅ + 0.05 K + 1 NAA	+	+++
2.	B₅ + 0.05 K + 2 NAA	+	+
3.	B₅ + 0.05 K + 5 NAA	+	++
4.	B₅ + 0.05 K + 1 NAA + 1, 2, 4-D	++	++
5.	B₅ + 0.05 K + 1 NAA + 2, 2, 4-D	+	++
6.	B₅ + 0.05 K + 1 NAA + 5, 2, 4-D	+	++

+ Callus initiated; ++ Normal callus formation; +++ Good callusing.

these plants were maintained in light they flowered profusely. When these plants were transferred in soil and gradually to the field, they showed healthy growth. (See also Tables 2A, 2B and 3, Fig. 4.)

DISCUSSION

The induction of callus in *Amaranthus paniculatus* can be made on different explants. The outer cells of the explants are involved in callus formation and an amorphous, creamy white callus is produced. In all the cases 2, 4-D was found to induce callus formation, but for growth of callus, other auxins such as IAA, NAA

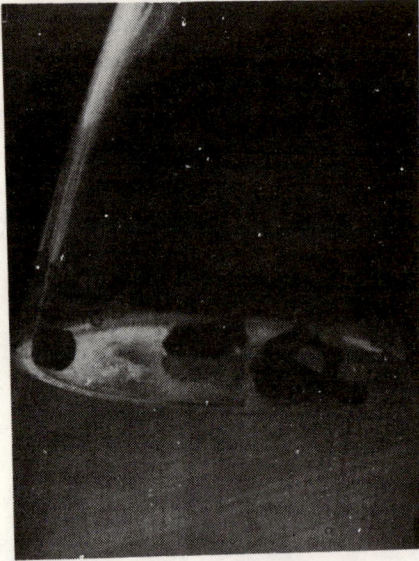

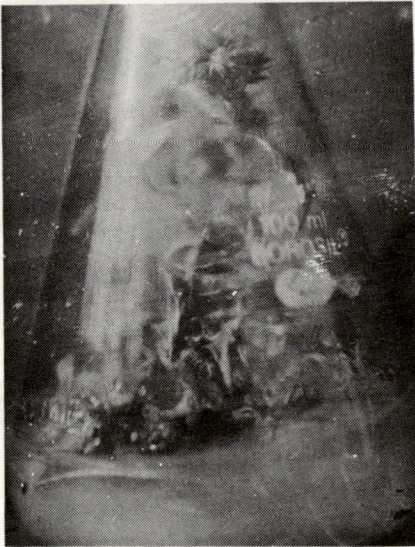

Fig. 1: Callus induction and growth on MS + NAA (2 mgl^{-1} + 2, 4-D (2 mgl^{-1}) + K (0.05 mgl^{-1}) medium.

Fig. 2: Shoot differentiation on MS + BAP (2 mgl^{-1}) + NAA (0.5 mgl^{-1}) containing medium.

Fig. 3: Culture plants on rooting medium (MS + 5 mgl^{-1} NAA).

Fig. 4: Plant transferred in pot.

Table 2A: Response of callus initiation and growth to varying auxin concentrations in MS medium

S. No.	MS medium supplemented with different concentration of auxin mg 1-^{1}	Cotyledon	Hypocotyl
1.	MS + 0.05 K + 1 NAA	++	++
2.	MS + 0.05 K + 2 NAA	++	+++
3.	MS + 0.05 K + 5 NAA	+++	+++
4.	MS + 0.05 K + 1 NAA	++	++
5.	MS + 0.05 K + 2 NAA	++	+++
6.	MS + 0.05 K + 5 NAA	+++	+++
7.	MS + 0.05 K + 1, 2, 4-D	++	+++
8.	MS + 0.05 K + 2, 2,4-D	+++	+++
9.	MS + 0.05 K + 5, 2, 4-D	++++	++++
10.	MS + 0.05 K + 1 NAA + 1, 2, 4-D	++	+++
11.	MS + 0.05 K + 1 NAA + 2, 2, 4-D	+++	+++
12.	MS + 0.05 K + 1 NAA + 5, 2, 4-D	++++	++++
13.	MS + 0.05 K + 5 NAA + 1, 2, 4-D	+++	+++
14.	MS + 0.05 K + 5 NAA + 2, 2, 4-D	++++	++++
15.	MS + 0.05 K + 5 NAA + 5, 2, 4-D	++++	++++
16.	MS + 0.05 K + 5 IBA + 2, 2, 4-D	++++	++++

+ Callus initiation; ++ Normal callusing;+++ Good callusing; ++++ Very good callusing.

Table 2-B: Response of callus initiation and growth to varying cytokinin concentrations in MS medium

S.No.	MS medium supplemented with different concentrations of cytokinins mg 1^{-1}	Cotyledon	Hypocotyl
1.	1 K + 0.5 NAA	++	++
2.	2 K + 0.5 NAA	++++	++++
3.	4 K + 0.5 NAA	+++	+++
4.	1 BAP + 0.5 NAA	++	+++
5.	2 BAP + 0.5 NAA	++++	++++
6.	4 BAP + 0.5 NAA	+++	+++
7.	1 K + 0.5 NAA	++	++
8.	2 K + 0.5 NAA	+++	++
9.	1 BAP + 0.5 NAA	++	++
10.	2 BAP + 0.5 NAA	++++	++++

+ Callus initiated; ++ Normal callusing; +++ Good callusing; ++++ Very good callusing.

Table 3: Differentiation response of callus and explants of seedling under different regime of auxins and cytokinins

S.No.	MS medium supplemented with different concentrations of Auxins + Cytokinins mg 1-¹	Callus		Hypocotyl & cotyledon	
		Shoot	Root	Shoot	Root
1.	MS + 1 NAA + 0.05 Kn	++	-	++	+
2.	MS + 2 NAA + 0.05 Kn	++	-	++	++++
3.	MS + 5 NAA + 0.05 Kn	++	-	++++	++
4.	MS + 10 NAA + 0.05 Kn	++	-	++++	++
5.	MS + 1 IAA + 0.05 Kn	++	-	+++	++
6.	MS + 2 IAA + 0.05 Kn	++	-	+++	+++
7.	MS + 5 IAA + 0.05 Kn	++	-	++++	+++
8.	MS + 10 IAA + 0.05 Kn	++	-	++++	++
9.	MS + 1, 2, 4-D + 0.05 Kn	-	-	-	-
10.	MS + 2, 2, 4-D + 0.05 Kn	-	-	-	-
11.	MS + 5, 2, 4-D + 0.05 Kn	-	-	-	-
12.	MS + 1, Kn + 0.05 NAA	+++	-	+++	-
13.	MS + 2Kn + 0.5 NAA	++++	+	++++	+
14.	MS + 1 BAP + 0.5 NAA	+++	-	+++	-
15.	MS + 2 BAP + 0.5 NAA	+++	-	+++	-
16.	MS + 1Kn + 0.5 IAA	+++	-	+++	+
17.	MS + 2 Kn + 0.5 IAA	++++	+	+++	-
18.	MS + 1 BAP + 0.5 IAA	+++	-	+++	-
19.	MS + 2 BAP + 0.5 IAA	+++	-	+++	-

– No differentiation; + Differentiation poor; ++ Differentiation normal; +++ Good differentiation; ++++ Very good differentiation.

and cytokinins are needed. Yeoman and Mitchell [7] concluded that nucleolus is the site of hormonal action while Zwar and Brown [9] demonstrated the accumulation of 2, 4-D in nucleoli. The exogeneous addition of phytohormones in media to induce callus has been reported by earlier workers also [8]. The regeneration of plants of *A. paniculatus* through callus and explants depends on exogenous application of hormones to balance the endogenous level. Regulation of root, shoot differentiation by auxin, cytokinin qualitative ratio is demonstrated by a number of workers [1]. Interestingly in the present studies both auxins and cytokinins induced shoot formation. Rooting was obtained when a high amount of auxins was taken, indicating that the endogenous level of auxins in cells is low. The formation of plants from callus and explants in *A. paniculatus* and their subsequent subculture is reported here for the first time. These plants flowered and survived when transferred to soil. These studies open the field for somaclone isolation of somatic variants which can be applied for crop improvement.

LITERATURE CITED

1. Flick, C.E., D.A. Evans and W.R. Sharp. 1983. Organogenesis. In: *Hand Book of Plant Cell Culture*, Vol. I edited by Evans, Sharp, Ammirato and Yamada. MacMillan Publ., New York.

2. Gamborg, O.L., P.A. Miller and K. Ojino. 1968. Nutrient requirements of suspension cultures of soybean root cells. *Exp. Cell Res.*, 50: 151-158.

3. Heller, R. 1953. Researches sur la nutrition minerale des tissus végétaux cultivars in vitro. *Ann. Sci. natl. Biol. Veg.* 14: 1-223.

4. Murashige, T. and F. Skoog. 1962. A revised medium for rapid growth and bioassays with tobacco tissue cultures. *Physiol. Plantarum.*, 18: 100-127.

5. Nitsch, J.P. and C. Nitsch. 1956. Auxin dependent growth of excised Helianthus tuberous tissues. *Am. J. Bot.* 43: 839-851.

6. White, P.R. 1963. *The cultivation of Animal and Plant Cells*. Ronald Press, New York, 2nd ed.

7. Yeoman, M.M. and J.P. Mitchell. 1970. Changes accompanying the addition of 2, 4-D to excised Jerusalem artichoke tuber tissue. *Ann. Bot.*, 34: 799-810.

8. Yeoman, M.M. and A.J. Macleod. 1977. Tissue (callus culture techniques). In: *Plant Tissue and Cell Culture*. edited by H.E. Street. Oxford: Blackwell Sci. Publ., 2nd ed. pp. 31-59.

9. Zwar, J.A. and R. Brown. 1968. Distribution of labelled plant growth regulators within cells. *Nature* (London) 220: 500-501.

13

Ecology and Agriculture

Mervyn W. Thenabadu

Faculty of Agriculture
University of Peradeniya
Peradeniya, Sri Lanka

INTRODUCTION

Ecology refers to the total system of interrelated elements existing in a specified area. The major components of ecology include the natural and physical habitat or environment together with the social and cultural systems. Environment refers to the aggregate of all external conditions and influences affecting the life of organisms. The social and cultural parameters of ecology are anchored basically in physical and natural habitats, such as islands, plains, valleys and, finally, the continents.

The ecological limitations of an area together with the materials and techniques at the disposal of the environment limit the agricultural practices of peasant farmers. Thus it is not possible to visualise a single model of a peasant cultivator. Therefore, it is best to concentrate on known examples from the SAARC region, in which the interest of this Regional Conference is centred.

Agriculture occupies, and will continue to occupy, a leading place in the economies of most SAARC countries. There is a complex variety of agricultural patterns on the earth's surface, many of which exist in our area of the world, which have traditional age-old beginnings.

A knowledge of the location of agricultural activity and production will enable us to predict the likely results of any proposed changes.

The SAARC countries are different from the Western, technologically advanced, more developed countries due primarily to the existence of large number of peasant farmers who operate on a small scale and are dependent on the environment. They benefit little from the few links they have with the external world and enjoy a relatively simple economic climate which, however, is subject to radical and rapid change.

CLIMATE

Climate or the environment has to satisfy two essential needs for the survival of plants. These are radiant energy and moisture. For a crop community these two parameters must be defined and known.

Both vegetation and land use are functions of physical and human environments, which are essentially multivariate.

Climate parameters of ecological significance are energy and moisture balances. Fogg, in 1986, referring especially to Africa, stated that each of the large number of places particularly involved might be a 'law unto itself' with respect to the conditions of the environment, which would necessarily increase the complexity.

The important environmental parameter, climate, greatly influences all facets of agricultural activity, be it the choice of a crop, crops or crop rotation and cultural practices, seedbed preparation, crop protection, crop nutrition or harvest and post-harvest operations. Even the introduction of new crops into a new area is dictated by climate to a very great extent.

Eighty percent of the rice-growing areas of south and south east Asia are rainfed and depend on monsoonal activity, which is characterised by high spatial and temporal variability. The time of planting is determined by the onset of the monsoon. The growth and development of the crop is much influenced by the subsequent temporal distribution of rainfall. Where farmers depend on supplemental irrigation, the filling of reservoirs after the monsoon ceases, will also depend on the preceding rains in a particular year.

Climatic or agro-climatic classification of areas is useful in planning cropping calandars, although they may suffer from limitations. Other criteria such as ratio of precipitation and temperature, moisture index, length of growing season, vapour pressure deficit water-balance models indicate agricultural potentialities of regions.

The link between the physical environment and man is seen when one considers man's food-getting economies. It is within this framework that men of various human groups appraise and turn to their own purpose the potentialities of the environment. The way the environment is exploited or used, and therefore the meaning of the environment to any human group is related to the culture of the group. An environment will have limited uses to a group if its technology is not developed. Such a group will possess only a limited range of resources if its technology is relatively primitive and undeveloped. For example, a primitive tribe in a tropical rain-forest will be dependent only on the wild animals and wild berries and fruits for their food. As technology advances, however, the meaning of the environment would also change, and the range of exploitable resources is therefore bound to be wider.

In the SAARC region we see a wide range of economies, ranging from a simple food-gathering economy to sophisticated urban-industrial economies in the cities. Between these extremes lie a spectra of agricultural economies such as shifting cultivation, sedentary cultivation of uplands, dry lands which are dependent on rains and sedentary cultivation of land dependent on highly developed irrigation techniques.

TYPES OF FARMING

Agriculture depends primarily on plants, upon which the food of all animals including humans depend. In addition plants provide fibre and materials for healing of ailments and diseases of man and animals. Even carnivorous animals depend on herbivorous animals, who in turn depend primarily on plants for food. Fortunately, there is much genetic variability in the SAARC region. Further, developed countries of the western hemisphere have successfully introduced valuable plants and even modified plant characteristics for favourable yields and adaptation to the new environments. In our countries people have adapted potentially promising varieties of crop plants through many decades or centuries of selection for characteristics they considered beneficial.

In this region we can identify a few basic types of farming, the most important of which are as follows:

1. Shifting cultivation
2. Home-garden cultivation
3. Dry-field cultivation, and
4. Wet-rice cultivation.

Each of these types is governed by ecological and environmental conditions and would fit into a particular ecological niche. These basic types of farming are discussed below.

1. Shifting Cultivation

Shifting cultivation is the system of forest/field-crop rotation agriculture, where the tree cover of the forest is allowed to regenerate and protect, and renew and replenish the fragile and vulnerable soil which is so characteristic of the tropics. In addition, the ash of the burnt forest enriches the fertility of the otherwise impoverished soil. The system may be referred to as one of rotation of fields rather than rotation of crops —a system which depends on the forest cover to maintain fertility in areas of low man/land ratios. The shifting cultivators of South-East Asia have blended themselves into the existing ecological equilibria of the climate-vegetation soil complex in their areas. In this they have created for themselves a way of living or an art of agriculture which disrupts the equity to a minimum because these agriculturists are not interested in mastering nature. They live in harmony with nature. It is probably because they are incapable of doing (mastering nature), that they have adopted a passive adjustment to the environment. Thus they differ from the rice-growing agriculturists who dominate their environment.

2. Home-garden Cultivation

This type of agriculture is a reflection of ecological and social conditions. Here peasants enjoy the security of tenure which encourages the growing of perennial plants, around their homes.

This garden cultivation of agriculture contrasts sharply with wet-rice cultivation in being essentially multi-storeyed with banana, papaya and tree crops integrated with a range of vegetable crops, dictated by the environment. Generally, it does not include grains or cereals and therefore is concerned with the production of crops which are not required in large quantities, but which are needed to supplement the staple diet and to add flavour to the basic meal of a cereal, such as rice.

This type of agriculture is a reflection of the ecological and social conditions, wherein peasants enjoy the security of tenure which encourages the growing of perennial plants in their own plots of land. If the land did not belong to them, it is quite unlikely that they would grow perennial plants that rarely bear produce for at least four or five years.

3. Dry-field Cultivation

Dry-field cultivation displays a tendency to spread beyond the areas of water control. Further, it shows a tendency to break away from shifting cultivation to one of sedentary agriculture. This type of agriculture leads to extensive deforestation and invariably to accelerated soil erosion. During the periods of European occupation in the SAARC areas, from the 16th century onwards, this type of agriculture or plantation agriculture, which exploited a few plants, was encouraged. This type of agriculture did not necessarily involved food plants.

One severe problem cultivators face with this type of farming is maintenance of soil fertility. The use of green manures and farm manures has been practised therefore with this type of cultivation of dry lands.

4. Wet-Rice Cultivation

Wet-rice cultivation occupies a distinctive ecological niche, the environment which Spencer termed, the 'worst aquatic fringe'. As facilities for irrigation developed, wet-rice culture expanded into certain ecologically different areas. Through trial and error, during long centuries, famers have selected many thousands of rice varieties with differing lengths of growing season (two to ten months) and a wide range of tolerance to soil and hydraulic conditions (salinity) and adverse environmental conditions. The farmers know which rice varieties are best suited to their environment and the best time of the year to begin and end their cropping calendars, which are synchronised with the weather patterns of the areas.

FUTURE TRENDS

Today, expansion in agricultural production has increased to an extent that the equilibrium in the environment is being threatened. Some big development projects must certainly have had their impact on an otherwise undisturbed environment. In addition, the extensive use of agrochemicals for promoting plant growth and development and for plant protection must also influence the delicately balanced ecology of the biosphere. Disposal of agricultural wastes is also a problem in certain parts of the world, where the valuable natural sources of ground-water and the atmosphere are easily polluted.

14

Proline Accumulation in Some Halophytes in the Indian Desert

David N. Sen and Sher Mohammed

Laboratory of Plant Ecology, Botany Department,
University of Jodhpur, Jodhpur 342001, India

ABSTRACT

The phenomenon of free proline accumulation in plants exposed to diverse environmental stresses has considerable eco-physiological significance. Proline accumulation occurs in the tissues of plants which grow in a saline habitat and acts as an endogenous osmotic regulant in halophytes. The present study revealed that all halophytes which grow in saline areas exhibited higher proline during cold stress at three selected sites in the Indian desert, viz., Pachpadra (site-I, extreme saline), Didwana (site-II, saline) and Jodhpur (site-III, non-saline). *Suaeda fruticosa, Salsola baryosma, Trianthema triquetra* and *Sporobolus helvolus* accumulated more proline at site-I than at the other two sites. Among all the plants, *Zygophyllum simplex* accumulated the maximum proline (109.9 µg/g fr. wt.). Since site-I is more saline, it can be concluded that salt stress caused plants to accumulate more proline and that perhaps it played some role in their survival.

INTRODUCTION

One of the few physiological differences demonstrated so far between halophytes and glycophytes is the products of dark fixation of carbon dioxide. In several species of halophytes, the major products of dark fixation are amino acids [1]. Although proline is normally a minor component of the pool of free amino acids in glycophytes, it has been observed to accumulate under conditions of water stress. The phenomenon of free proline accumulation in plants exposed to diverse environmental stresses has considerable eco-physiological significance. Proline is a five-carbon amino acid which accumulates in leaves under stress. Proline has also been known to accumulate in the leaves of many plant species when subjected to low temperature, salt stress or even starvation [2, 3]. Recently, it was also observed that proline is not necessarily an indicator of a stressed condition [4]. It was found that proline accumulation occurs in the tissues of plants exposed to a saline substrate [5]. It was also observed that proline accumulation was not governed by environmental factors; rather, some innate factors were responsible for it [6]. It has been suggested that proline acts as an endogenous osmotic regulant in halophytes. Saline stress induces proline accumulation in a similar manner in plants subjected to water stress [7]. In halophytes, a positive correlation was seen between the proline content and the amount of $Na^+ + Cl^-$ in cell sap [8].

The relationship between proline accumulation and stress (drought) resistance raises the possibility that proline may also be implicated in resistance to physiological drought under saline conditions. The present study was thus directed towards obtaining a better understanding of the proline level in some halophytic species, growing at three different sites. The role of proline in halophytes is proposed.

MATERIALS AND METHODS

Plants were collected from nature at different sites in the Indian desert during 1985-87. Eight species, both annuals as well as perennials, were selected for the present investigation. Of these eight, three species were common to all the three sites (Pachpadra-I; Didwana-II; Jodhpur-III), viz., *Suaeda fruticosa*(L.) Forsk. (Fig. 1), *Sesuvium sesuvioides* (Fenzl.) Verdc. (Fig. 2) and *Trianthema triquetra* Rottler ex Willd. (Fig. 3) *Cressa cretica* Linn. (Fig. 4) and *Sporobolus helvolus* (Trin.) Th. Dur. et Schinz (Fig. 5) were found only at sites-I and II, while *Salsola baryosma* (Roem. et Schult.) Dandy (Fig. 7) at sites-I and III. *Aeluropus lagopoides* (Linn.) Trin. ex. Thw. Enum. (Fig. 6) and *Zygophyllum simplex* Linn. (Fig. 8) were observed only at site-I.

Fully mature leaves were collected in the morning and the estimation of proline was carried out in triplicate following the method of Bates and co-authors [9]. The osmotic potential of the leaf cell sap was measured as per the method of Janardhan and co-authores [10].

The obtained data were statistically analysed after Gomez and Gomez [11].

The climate is arid with an average annual precipitation of 128.0, 221.0 and 110.2 mm at sites-I, II and III respectively, thus all the sites could be included in the arid zone. The effective precipitation is received mainly during the monsoon.

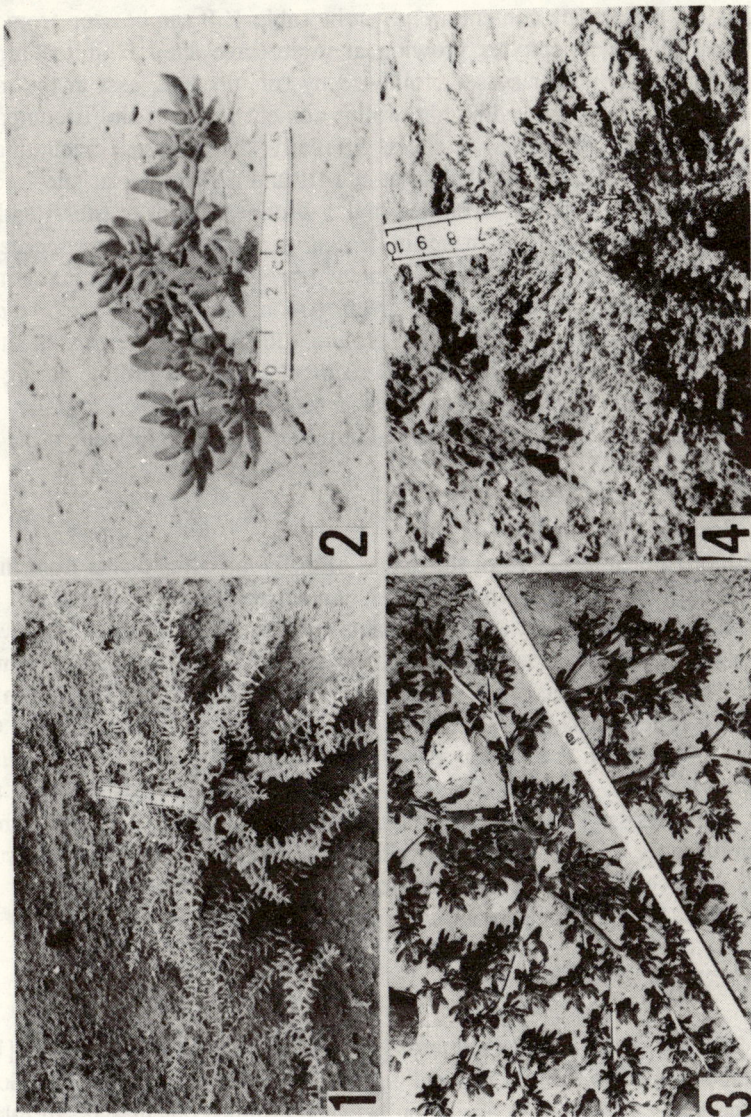

Figs. 1-4: Salt-accumulating fleshy plants *S. fruticosa* (1), *S. sesuvioides* (2), *T. triquetra* (3) and salt-secreting plant *C. cretica* (4) growing under field conditions.

RESULTS

Proline Content

Data on the proline content are given in Table 1. It can be seen trom this Table that all the halophytes growing at these three sites exhibited higher proline during the winter season, followed by summer and least in the rainy season. *S. fruticosa* showed higher proline content at sites-I and II compared to site III, being maximum at site-I. Similarly, *S. baryosma* accumulated slightly higher proline at site-I than at site-III and *S. helvolus* at site-I compared to site-II. *T. triquetra* exhibited higher proline content at sites-I and III than at site-II, being maximum at site-I. *A. lagopoides* and *C. cretica* had a higher amount at site-I than site-II. Of the two strains of *Z. simplex*, red strains exhibited more proline compared to green. Among all the plants, *Z. simplex* (red strain) accumulated the maximum proline, followed by *C. cretica* at site-I. Comparatively, all plant species of site-I accumulated higher proline during the winter season, followed by site-II and least at site-III.

The data obtained on proline from all three sites were significant at the 5% level.

Plant Osmotic Potential

Osmotic potentials (OP) of the leaves of different saline plants from three different sites are presented in Table 2. It can be seen from this Table that the higher values of OP (-bars) in all the plants collected from all three sites were obtained in the rainy season, followed by summer and least in the winter season. The highest value of OP (-bars) was observed in *S. sesuvioides* from all three sites, followed by *S. helvolus* and *A. lagopoides*. In *S. baryosma* the OP (-bars) was lowest in the winter season, followed by *S. fruticosa*, *C. cretica*, *T. triquetra* and *Z. simplex*. These low values (-bars) in leaves during the winter season, possibly due to continuous uptake of soluble ions, create a stress condition for the species. They maintain a high concentration of osmotic substances to successfully overcome stress conditions, of which proline is one.

The data from all three sites were significant at the 5% level.

DISCUSSION

Water and salt stress induces numerous metabolic irregularities in plants [12, 13]. Plants treated to salinity showed an increased level of proline [14]. Goas [15] reported that there is a correlation between proline content and salt tolerance. The physiological significance of proline accumulation under stress is not known. Proline at high concentrations acts as a source of solute for intracellular osmotic adjustment [16], a storage compound for both nitrogen and carbon for utilisation in growth after stress [17] and as a protective agent to the enzymes and cellular structures [18]. However, Chu and co-authors [3] sus-

Table 1: Seasonal variations in proline (μg/g fr. wt.) content in halophytes growing at three sites (I-III)

Species	Pachpadra (site-I)			CD at 5%	Didwana (site II)			CD at 5%	Jodhpur (site-III)			C at 5%
	Rainy	Winter	Summer		Rainy	Winter	Summer		Rainy	Winter	Summer	
Aeluropus lagopoides	6.21	26.65	-	4.17	24.63	-	-	a	-	-	-	-
Cressa cretica	0.51	108.03	3.83	26.41	4.79	-	-	a	-	-	--	-
Salsola baryosma	0.09	7.12	5.75	0.87	-	-	-	-	0.14	7.05	2.74	2.41
Sesuvium sesuvioides	1.99	-	-	a	4.35	-	-	a	1.80	-	-	a
Sporobolus helvolus	5.33	99.99	7.90	27.55	6.15	8.70	6.52	0.74	1.90	-	-	-
Suaeda fruticosa	1.40	11.93	7.02	0.82	2.68	9.86	6.57	8.66	1.90	5.76	5.00	2.39
Trianthema triquetra	3.38	74.73	-	17.98	3.75	14.33	-	12.34	5.23	15.56	8.16	1.30
Zygophyllum simplex (G)	6.36	64.18	-	17.83	-	-	-	-	-	-	-	-
Z. simplex (R)	5.67	109.93	-	42.30	-	-	-	-	-	-	-	-

G = green strain; R = Red strain; -- = plant not seen; a = data insufficient for analysis.

Figs. 5-8:   Salt-secreting grasses *S. helvolus* (5). *A. lagopoides* (6) and salt-accumulating fleshy plants *S. baryosma* (7) and *Z. simplex* (8) growing under field conditions.

Table 2: Seasonal variations in osmotic potential (-bars) of halophytes growing at three sites (I-III).

Species	Pachpadra (site-I)			CD at 5%	Didwana (site-II)			CD at 5%	Jodhpur (site-III)			CD at 5%
	Rainy	Winter	Summer		Rainy	Winter	Summer		Rainy	Winter	Summer	
Aeluropus lagopoides	33.55	49.65	-	6.29	29.45	-	-	a	-	-	-	-
Cressa cretica	43.88	70.00	85.00	5.10	51.32	-	-	a	-	-	-	-
Salsola baryosma	61.00	107.40	105.00	5.92	-	-	-	-	70.00	120.01	89.85	41.96
Sesuvium sesuvioides	19.59	-	-	a	13.47	-	-	a	42.00	-	-	-
Sporobolus helvolus	21.90	54.43	41.03	13.38	28.74	59.34	53.51	15.92	55.00	78.82	16.91	66.77
Suaeda fruticosa	59.69	91.52	89.50	15.59	58.72	85.41	84.49	17.80	52.09	58.34	55.00	0.92
Trianthema triquetra	32.57	69.47	-	27.22	29.47	65.32	-	49.76	-	-	-	-
Zygophyllum simplex (G)	30.42	55.44	-	14.32	-	-	-	-	-	-	-	-
Z. simplex (R)	41.03	69.08	-	41.08	-	-	-	-	-	-	-	-

G = Green strain; R = Red strain; - = plants not seen; a = data insufficient for analysis.

pected the potentiality of proline to bring about osmotic adjustments under saline conditions. Adaptive features, such as increase in proline level, develop during severe conditions but these are to be related more to survival than to performance [19].

The present investigation deals with the relationship between proline accumulation and osmotic potential in leaves of halophytic plant species. All halophytic species from three different sites showed low (-bars) osmotic potential (OP) in their cell sap during the winter season, followed by summer and highest in the rainy season. Furthermore, all plant species also accumulated more proline during the same season (winter), followed by summer and least in rainy. So, it can be said that during an unfavourable period, soil-water could not be absorbed by the plants due to the low water potential of the medium or due to continuous uptake of soluble ions which create a stress condition in plant cells. Salts in the medium generally decreased the water availability to the roots, thus creating a situation of physiological drought. Thus accumulation of proline acts as an intra-cellular osmotic adjustment in their cell sap and it is quite probable that proline may be playing some role for tolerance of harsh saline conditions. Osmotic adjustment or osmoregulation enables plants to maintain growth as plant water potential decreases. Adjustment occurs through decrease in osmotic potential by solute accumulation in cells as leaf water potential decreases. So, the accumulation of high concentrations of proline in the leaves of halophytes under saline conditions suggests that it might have a role in osmotic adjustment. Certainly proline meets the requirements of a compound with this function because it has a high solubility; it is a neutral compound and at high concentrations has little effect on enzyme activity [16]. In the present investigation, all plant species at site-I accumulated more proline and exhibited low (-bars) osmotic potential (OP) compared to sites-II and III, which may be due to the high salinity of this habitat. So, it can be concluded that salt stress caused the plants to accumulate more proline. Perhaps free proline contents play an essential role in their survival and osmotic adjustment.

Acknowledgements

We are thankful to the Head of Botany Department for facilities. Financial assistance received from the Department of Environment, New Delhi, is also acknowledged.

LITERATURE CITED

1. Webb, X.L. and J.W.A. Burley. 1965. Dark fixation of CO_2 by obligate and facultative salt marsh halophytes.*Can. J. Bot.*, **48**: 281-285.
2. Aspinall, D., T.N. Singh and L.G. Paleg. 1973. Stress metabolism. V. Abscisic acid and nitrogen metabolism in barley and *Lolium temulentum* L. *Aust. J. Biol. Sci.*, **26**: 319-327.
3. Chu, T.M., D. Aspinall and L.G. Paleg. 1976. Stress metabolism. VII. Salinity and proline accumulation in barley. *Aust. J. Pl. Physiol.*, **3**: 219-228.

4. Mohammed, S. and D.N. Sen. 1987. Proline accumulation in arid zone plants. *Jour. Arid Environ.*, **13**: 231-236.

5. Palfi, G. and J. Juhasz. 1970. Increase of free proline level in the water deficient leaves as a reaction to saline or cold root media. *Acta. Agron. Acad. Sci. Hung.*, **19**: 278-281.

6. Mohammed, S. and D.N. Sen. 1990. Environmental changes and proline content in some desert plants. *Jour. Arid Envison.*, **19**: 241-243.

7. Singh, T.N., D. Aspinall, and L.G. Paleg. 1972. Proline accumulation and varietal adaptability to drought in barley; a potential metabolic measure of drought resistance. *Nature* **236**: 188-190.

8. Triechel, S. 1975. Der Einfluss von NaCl auf die Prolinekon zentration verschiedener Halophyten. *Z. Pflanzenphysoil.*, **76**: 56-58.

9. Bates, L.S., R.P. Waldren and I.D. Teare. 1973. Rapid determination of free proline for water stress studies. *Plant & Soil* **39**: 205-207.

10. Janardhan, K.V, A.S.P. Murthy, K. Giriraj and S. Panchaksharaih. 1975. A rapid method for determination of osmotic potential of plant cell sap. *Curr. Sci.*, **44**: 390-391.

11. Gomez, K.A. and A.A. Gomez. 1984. *Statistical Procedures for Agricultural Research* John Wiley & Sons, New York, USA. (2nd ed.).

12. Levitt, J. 1972. *Responses of Plants to Environmental Stress.* Academic Press, New York.

13. Chu, T.M., D. Aspinall and L.G. Paleg. 1974. Stress metabolism. VI. Temperature stress and accumulation of proline in barley and radish. *Aust. J. Pl. Physiol.*, 1: 87-99.

14. Wyn Jones, R.G. and R. Storey. 1978. Salt stress and comparative physiology in Graminae. II. Glycine betaine and proline accumulation in two salt and water stressed barley cultivars. *Aust. J. Pl. Physiol.*, **15**: 817-829.

15. Goas, M. 1971. Metabolism axote des halophytes. Utilization de la L-proline 14^c (U) par les jeunes plates d'*Aster tripolium L. C. R. Acad.* (Paris), **272**: 414-417.

16. Stewart, G.R. and J.A. Lee. 1974. The role of proline accumulation in halophytes. *Planta*, **120**: 279-289.

17. Barnett, N.M. and A.W. Naylor. 1966. Amino acids and protein metabolism in *Bermuda* grass during water stress. *Plant Physiol.*, **41**: 1222-1230.

18. Schobert, B. and H. Tschesche. 1978. Unusual properties of proline and its interaction with proteins. *Biochem. Biophys. Acta*, **541**: 270-277.

19. Greenway, H. and R. Munns. 1980. Mechanism of salt tolerance in nonhalophytes. *Ann. Rev. Pl. Physiol.*, **31**: 149-190.

15

Ecophysiological Effects of Some Heavy Metallic Pollutants on Two Aquatic Macrophytes

Govind Ghimire* and K.M.M. Dakshini

Dept. of Botany, University of Delhi, Delhi, India

ABSTRACT

Two aquatic macrophytes—*Alternanthera philoxeroides* and *Ludwigia adscendens* —were introduced into different kinds of aquatic media, viz., water from the Yamuna River, Hindon River and major drains of the Union Territory of Delhi.

The effects of metallic pollutants on plants were judged by these parameters: length of shoot tip, internode growth, root initiation and root growth.

The above aquatic media were treated with different concentrations of toxic metals, e.g., copper, lead, zinc and chromium. It was found that the uptake of metals correlated with their concentration in the medium; the preference or capability of plant uptake of metal ions varied, with Zn coming first, followed by Cu and Pb. The antagonistic effects of metal ions recorded in this study, especially the inhibition of root growth, indicate that these two aquatic macrophytes respond to metallic toxicity. Thus *A. philoxeroides* and *L. adscendens* could serve as bio-indicators and bio-monitors of metallic pollution of water.

INTRODUCTION

The physiological effects of heavy metals as trace elements in plants are noted since quite some years ago [1]. Their behaviour as heavy metal pollutants on aquatic macrophytes has recently attracted many scientists—Bhargava [2], Mudroch and Capobianco [3] and others.

* Present address: Dept. of Botany, T.U. Kirtipur, Kathmandu, Nepal.

The present study sought to assess the morphological and chemical responses of aquatic macrophytes (*Ludwigia adscendens* and *Alternanthera philoxeroides*) to pollution caused by Cu, Pb, Zn and Cr as heavy metal pollutants. As the study progressed it became evident that the morphological responses of these two aquatic macrophytes could make them useful as indicators and monitors of heavy metal pollution since the parameters of deleterious effects are openly manifested and hence readily assessable.

Thus the main objective of this study became tagging (if possible) these two aquatic macrophytes as Water Quality Indicators and Monitors for the heavy metals mentioned earlier.

METHODS

The two macrophytes and river waters were collected from the Yamuna (Y_1-Y_4) and Hindon (H_1 and H_2); drain waters (N_1-N_7) were also sampled (Fig. 1). These macrophytes were transplanted in garden ponds and 25 cm long twigs were used in all the experiments.

To meet the objectives of this study, 5 different sets of experiments were undertaken with seasonal monitoring.

Besides analysing the concentrations of heavy metallic pollutants in the experimental twigs, data for such simple morphological parameters as (1) length of shoot tip; (2) length of the third internode; and (3) leaf area were also recorded.

Expt. I was conducted to study the effect of river waters and drains on the length of shoot tip, internode length and leaf area while Expt. II was done to determine the effect of distilled water and river water dilutions of drain waters on concentrations of heavy metallic pollutants and thus on root growth and length of twigs. Expt. III was initiated to study the effect of different concentrations of various heavy metallic pollutants on root initiation and root growth. Expt. IV was done to ascertain the effect of distilled water and river water solutions of heavy metallic pollutants on root growth, plant height and heavy metallic pollutant content composition.

OBSERVATIONS

The figures 1–5 and Table 1 reveal the findings of the experiments that were carried out during 1983-86.

DISCUSSION AND CONCLUSIONS

Findings from Expt. I suggest that *A. philoxeroides* favoured higher concentrations of heavy metals than *L. adscendens* [4, 5].

Expt. II further supports Expt. I that *A. philoxeroides* is better equipped for

Table 1: Heavy metal concentrations (μgg^{-1}), length of shoot tip (cm), internode length (cm) and leaf area (cm²) of *Alternanthera philoxeroides* and *Ludwigia adscendens* twigs in waters of the Yamuna River (Y_1-Y_4), Hindon River (H_1-H_2) and drain water (N_1-N_7).

Sl No.	Water samples	A. philoxeroides Heavy metals (μgg^{-1}) Cu	Pb	Zn	Length of shoot tip (cm)	Inter-node length (cm)	Leaf area (cm²)	L. adscendens Heavy metals (μgg^{-1}) Cu	Pb	Zn	Length of shoot tip cm	Inter-node (cm)	Leaf area (cm²)
1.	Distilled water controlled	5.3 ± 0.2	7.0 ± 0.6	225.1 ± 5.0	0.9 ± 0.4	0.5 ± 0.2	0.1 ± 0.3	19.0 ± 2.0	13.7 ± 1.2	109.7 ± 0.6	6.1 ± 0.6	0.4 ± 0.0	0.1 ± 0.1
2.	Y_1	5.0 ± 1.0	7.0 ± 1.0	182.0 ± 5.1	0.1 ± 0.2	0.3 ± 0.0	0.6 ± 0.1	15.0 ± 0.0	2.3 ± 0.6	132.0 ± 2.0	4.7 ± 4.0	0.1 ± 0.0	0.02 ± 0.01
3.	Y_2	2.7 ± 0.6	4.3 ± 0.6	165.0 ± 1.0	2.4 ± 1.3	0.5 ± 0.1	0.1 ± 0.1	12.0 ± 1.0	12.7 ± 0.6	153.0 ± 1.0	4.5 ± 1.0	0.2 ± 0	0.01 ± 0.01
4.	Y_3	5.3 ± 1.2	3.0 ± 1.0	166.7 ± 1.2	4.5 ± 0.3	1.1 ± 0.6	0.1 ± 0.1	17.7 ± 0.6	7.0 ± 1.0	171.3 ± 2.3	2.6 ± 0.5	0.7 ± 0.4	0.03 ± 0.01
5.	Y_4	1.0 ± 0.6	14.7 ± 0.6	182.7 ± 4.6	5.4 ± 1.2	1.4 ± 0.8	0.02 ± 0.0	16.7 ± 0.6	5.7 ± 0.6	154.3 ± 0.6	3.5 ± 0.8	0.4 ± 0.3	0.03 ± 0.02
6.	H_1	2.3 ± 0.6	3.0 ± 1.0	136.0 ± 2.6	1.3 ± 0.8	0.2 ± 0.2	0.2 ± 0.0	12.7 ± 1.5	7.7 ± 0.6	176.7 ± 0.6	2.6 ± 0.3	0.7 ± 0.3	0.03 ± 0.03
7.	H_2	5.3 ± 0.6	6.7 ± 2.1	125.0 ± 0	2.1 ± 2.0	0.8 ± 0.4	0.1 ± 0.0	43.0 ± 2.0	15.0 ± 1.0	201.3 ± 1.5	0.3 ± 0.2	0.1 ± 0.0	0.08 ± 0.02
8.	N_1	4.3 ± 0.6	7.0 ± 1.0	231.7 ± 1.5	5.8 ± 1.2	1.7 ± 0.5	0.1 ± 0.1	14.0 ± 1.0	18.3 ± 0.6	143.0 ± 2.6	8.4 ± 2.0	1.0 ± 0.3	0.03 ± 0.01
9.	N_2	4.7 ± 0.6	5.7 ± 0.6	158.7 ± 3.5	5.7 ± 2.2	1.7 ± 1.1	0.2 ± 0.1	17.0 ± 1.0	3.7 ± 0.6	159.0 ± 1.0	5.0 ± 1.5	0.5 ± 0.2	0.08 ± 0.04
10.	N_3	5.7 ± 0.6	6.0 ± 0	162.3 ± 3.1	0.5 ± 0.0	0.4 ± 0.0	0.1 ± 0.1	15.3 ± 0.6	14.0 ± 1.0	231.3 ± 1.2	6.9 ± 0.5	0.9 ± 0.2	0.2 ± 0.1
11.	N_4	7.3 ± 0.6	4.7 ± 1.5	207.0 ± 5.0	1.3 ± 1.0	0.8 ± 0.2	0.03 ± 0.1	14.7 ± 1.2	6.3 ± 0.6	232.7 ± 2.5	9.5 ± 0.9	1.9 ± 1.0	0.07 ± 0.02
12.	N_5	9.7 ± 0.6	6.0 ± 1.7	192.0 ± 1.0	4.5 ± 3.0	1.7 ± 0.7	0.1 ± 0.1	44.0 ± 2.0	6.3 ± 0.6	212.7 ± 5.0	6.8 ± 0.2	1.4 ± 0.2	0.2 ± 0.2
13.	N_6	3.1 ± 1.0	1.3 ± 0.6	182.7 ± 1.2	1.1 ± 0.3	0.5 ± 0.4	0.1 ± 0.1	11.7 ± 1.5	2.7 ± 1.2	140.0 ± 2.0	5.5 ± 0.5	0.4 ± 0.3	0.1 ± 0.1
14.	N_7	4.7 ± 1.5	7.7 ± 0.6	291.7 ± 1.2	4.1 ± 3.2	1.2 ± 0.4	0.1 ± 0.3	12.0 ± 2.5	2.5 ± 1.5	192.0 ± 6.5	4.6 ± 2.2	0.1 ± 0.0	0.2 ± 0.1

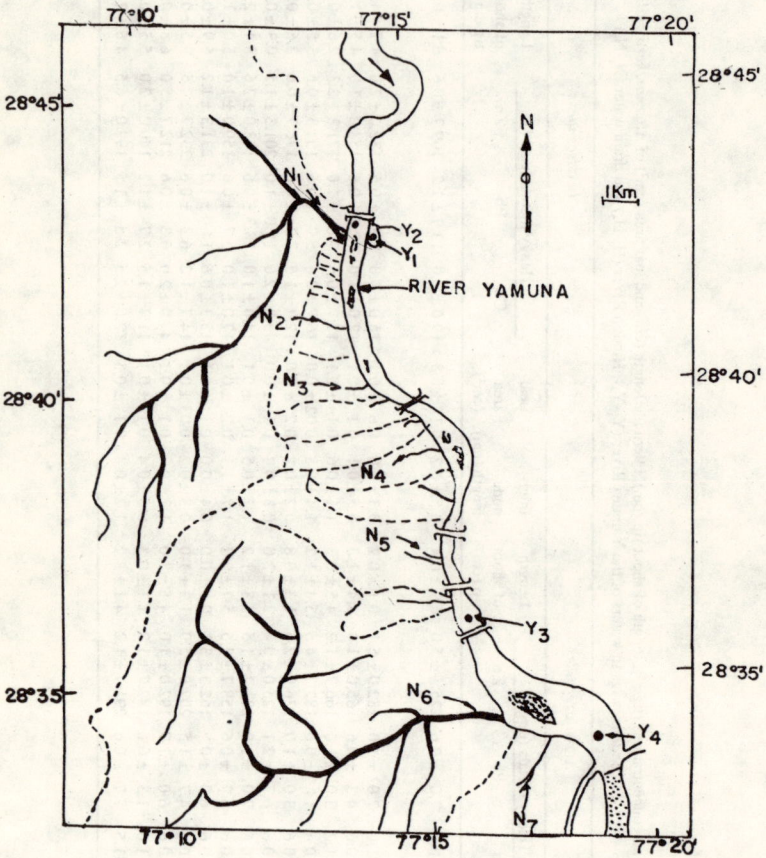

Fig. 1: Sketch map of the Yamuna River in the Union Territory of Delhi showing locations of sampling sites (Y_1-Y_4), important drains (N_1-N_7) and their basin areas (broken lines).

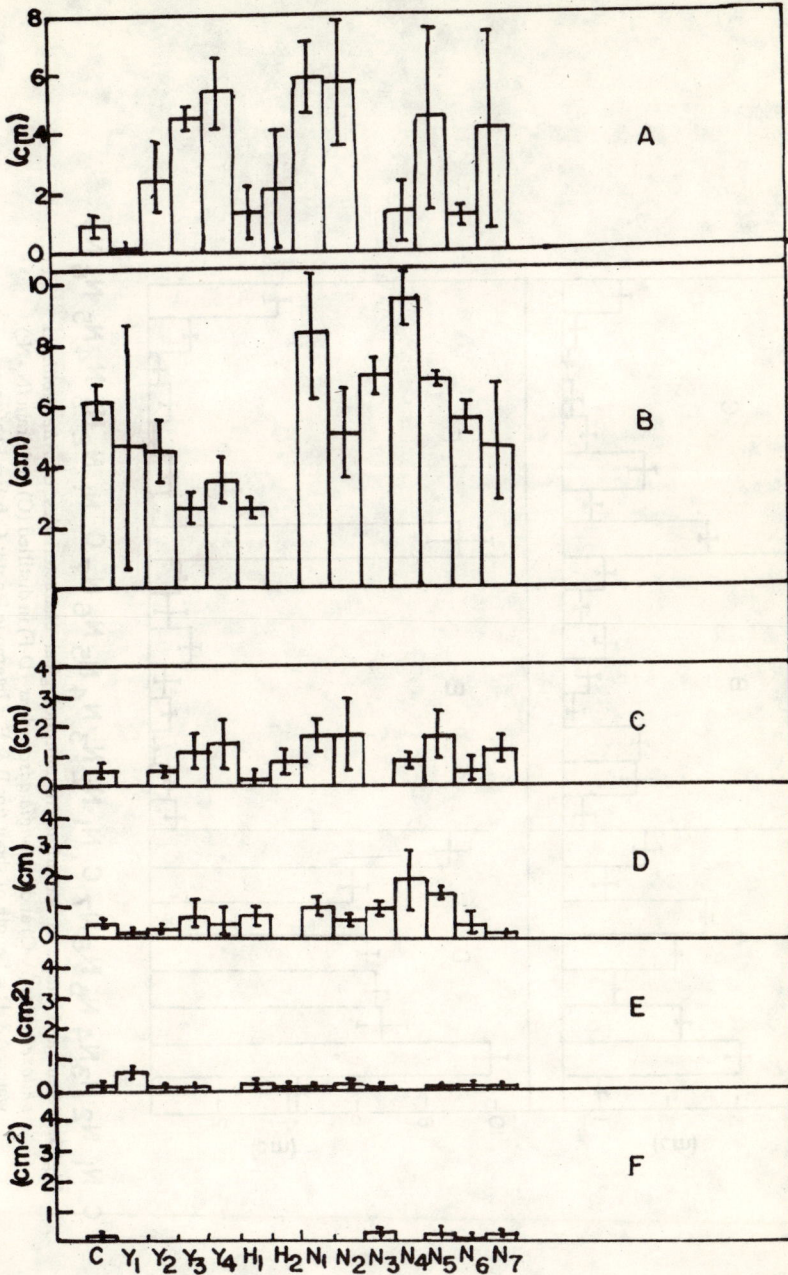

Fig. 2: Mean and range of variations in morphological characteristics of *Alternanthera philo-xeroides* (A,C & E) and *Ludwigia adscendens* (B,D & F) in distilled (C), river (Y₁-Y₄; H₁ and H₂) and drain (N₁-N₇) waters. A-B = Length of shoot tip; C-D = Internode length; E-F = Leaf area.

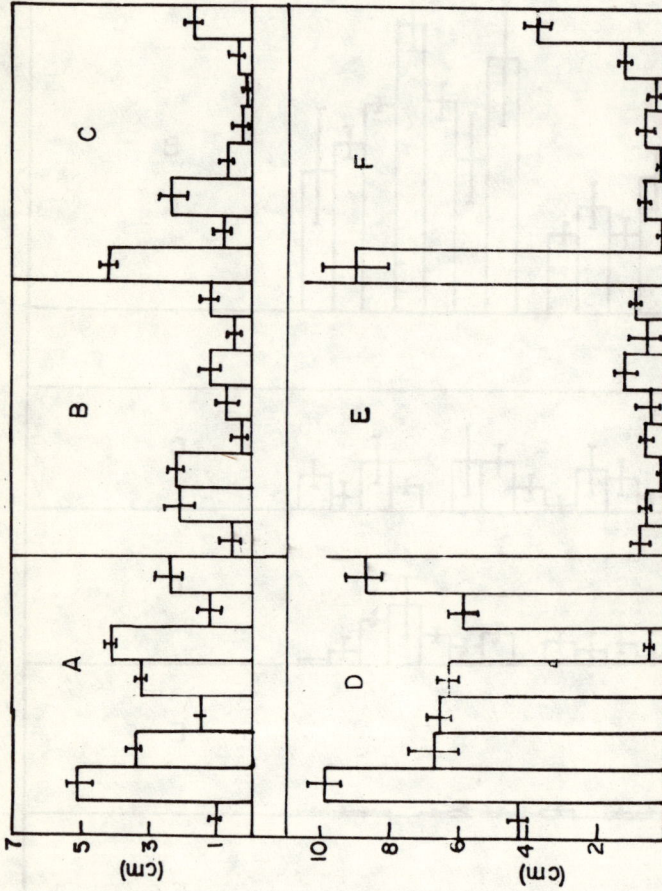

Fig. 3:   Mean and range of variations in morphological characteristics of *Alternanthera philoxeroides* (A-C) and *Ludwigia adscendens* (D-F) in distilled (C) and drain (N_1-N_7) water. A & D = Length of shoot tip; B & E = Internode length; C & F = Leaf area.

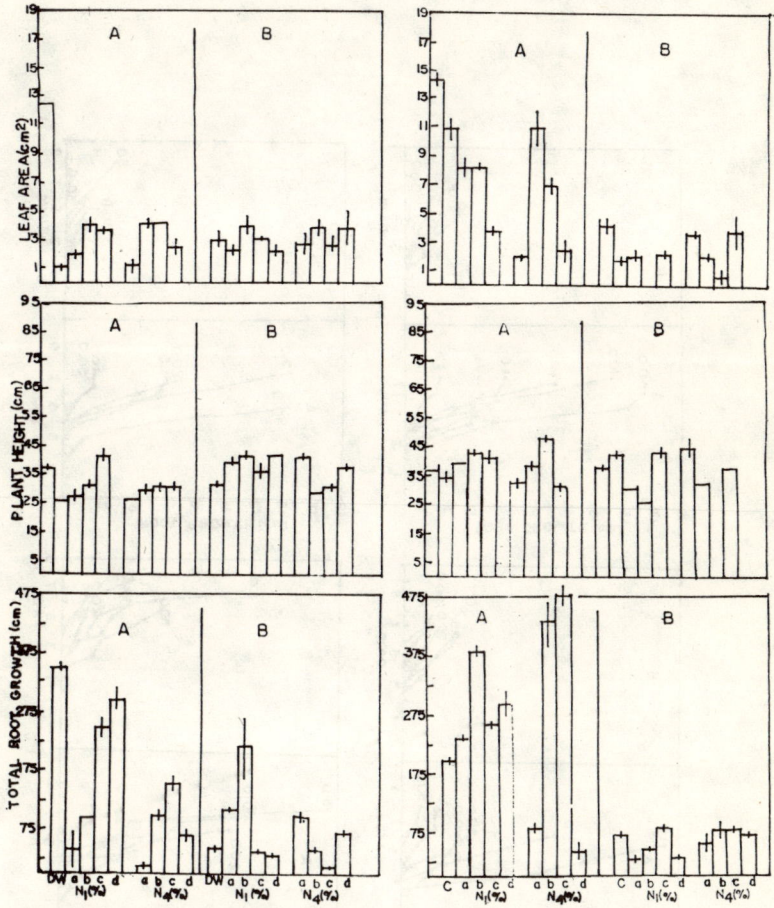

Fig. 4: Mean and range of variations in morphological characteristics of *Alternanthera philoxeroides* (A) and *Ludwigia adscendens* (B) in distilled water (DW) and river water (C) dilutions (25-100%) of N_1 and N_4 water.

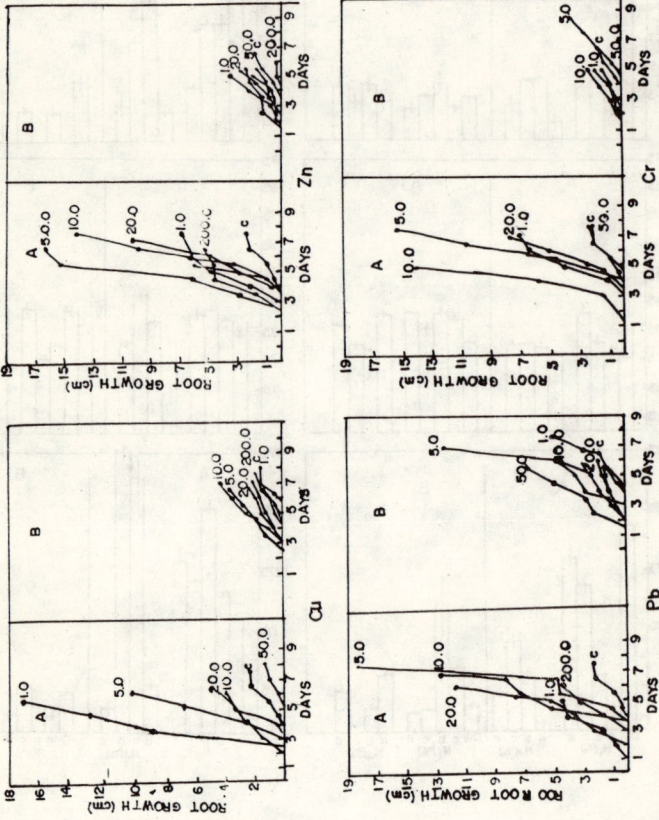

Fig. 5: Root initiation and growth in *Alternanthera philoxeroides* (A) and *Ludwigia adscendens* (B) in different concentrations (1-200 µgL⁻¹) of Cu, Pb, Zn and Cr.

higher concentrations of metal pollution than *L. adscendens*. Contrary to *L. adscendens*, *A. philoxeroides* grew better in drain water than in river water. This supports the importance of a natural medium for analysing the effect of metal pollutants over a medium of distilled water. Studies by Whitton [6] and Varga and Falls [7] support this conclusion.

Of the three morphological characters studied, length of shoot tip and internode length proved more responsive. This is more effective than the usual procedures of studying the nutrient deficiency symptoms, as supported by Cladwell [8], Davis and Lucas [9], Reuther and Labanauskas [10]. Roots were found sensitive [11] to the variations in cations in their vicinity and in fact such responses have been used for assessing the tolerance index of plants to various metallic pollutants [12]. Expt. III showed that root initiation commenced late in *A. philoxeroides* compared to *L. adscendens*, but the growth of the former was more rapid and more intense with increasing concentrations of heavy metals in water samples. In *L. adscendens*, the root growth was not only slow, but also exhibited no special preference for any concentration or for any metal in the water sample.

Data from Expt. IV confirmed a close parallel between the concentration of the various metals in the medium and the amount of each metal accumulated by the plants. With an increase in metal the concentrations in the medium, the relative preference, for them in the order of Zn, Cu and Pb likewise increased, with Cu at higher concentrations being an exception. It should also be noted that compared to distilled water solutions, in river water solutions the values of both these parameters were higher in *A. philoxeroides* than in *L. adscendens*. These observations demonstrate the extreme sensitivity of these plants to minor variations in metal concentrations in the medium.

This study thus highlights the fact that simple morphological features such as length of shoot tip, internode length, root growth and root index are of immense value for monitoring water quality. The significance of using natural water under laboratory conditions for a meaningful assessment of toxic load in an aquatic system has likewise been shown.

Another interesting outcome of the present study is the fact that as *A. philoxeroides* and *L. adscendens* respond differently to an ecological situation, the two plants can be used jointly as more reliable water quality indicators for heavy metals.

The water quality monitoring plants must be assessed, however, by growing them in their natural medium to obtain better and more natural results of the impact of pollutants on the biological as well as ecological systems as a whole.

In brief, therefore, these experiments, not only confirm the relative uptake of different heavy metallic pollutants by the two macrophytes, but also their responses to diverse ecological situations, exhibited by a wide gamut of characters encompassing simple morphological features as well as the nature of the heavy metal uptake mechanism.

Acknowledgements

We express our due regards to Head of the Botany Dept., Univ. of Delhi, Delhi for providing vehicles and other facilities.

LITERATURE CITED

1. Hoagland, D.R., A.R. Davis and P.L. Hibbard. 1948. The influence of one ion on the accumulation of another by plant cells with special reference to experiments with *Nitella. Pl. Physiol.*, 3: 473-486.
2. Bhargava, D.S. 1985. Water quality variation and control technology of Yamuna River. *Environmental Pollution* (Ser. A) 37: 355-376.
3. Mudroch, A. and J.A. Capobianco. 1979. Effects of mine effluent of uptake of Co, Ni, Cu, As, Zn, Cd and Pb by aquatic macrophytes. *Hydrobiologia*, 64: 223-231.
4. Anon. 1982. E.P.A.'s water quality criteria under attack. *Chem. Engng. News*, 29-32.
5. Rubin, A.B. 1981. Site-specific criteria modification—a method for field derivation of criteria to protect aquatic life. Symposium on Aquatic Toxicology, 6th American Society for Testing and Materials, October 13-14, St. Louis, MO.
6. Whitton, B.A.1975. *Studies in Ecology*, Vol. 2, *River Ecology*. Blackwell Scientific Publications, Oxford/London, 725 pp.
7. Varga, L.P. and C.P. Falls. 1972. Continuous system models of oxygen depletion in a eutropic reservoir, *Environ. Sci. Technol.*, 6: 135-142.
8. Cladwell, T.H. 1971. Copper deficiency in crops. In: Trace elements in soils and crops. *Min. Ag. Fish. Fd. Tech. Bull.* 21, HMSO, London, pp. 62-87.
9. Davis. J.F. and R.E. Lucas. 1959. Organic Soils, Their Formation, Distribution Utilization and Management. *Mich. Agric. Exp. Stn. Spec. Bull.*, 425 pp.
10. Reuther, W. and C.K. Labenauskas. 1966. Copper. In: Diagnostic Criteria for Plants and Soils, edited by H.D. Chapman. University of California, Riverside, pp. 157-179.
11. Bradshaw, A.D. and T. McNeilly. 1981. *Evolution and Pollution*. Institute of Biology's studies in Biology No. 130 London, Arnold, 76 pp.
12. Veltrup, W. 1977. The uptake of copper by barley roots in the process of zinc. *Z. Pflanzenphysiol.*, 83: 201-205.

16

Probable Direction of Evolution in the Genus *Anemone* L. *(Ranunculaceae)*

Ram P. Chaudhary

**Central Department of Botany, Tribhuvan University,
Kirtipur, Kathmandu, Nepal**

ABSTRACT

Altogether 14 species, 19 subspecies and 3 varieties of *Anemone* L. found in Nepal and adjacent regions of the Himalayas and Western China are disposed into 6 sections and 4 series. Based on comparative study of rhizome, leaf, inflorescence, nutlet and pollen grains, relative level of specialization of the sections and their phylogenetic relationship are proposed. Sections *Sylvia* Spach and *Rivularidium* Jancz. possess highly specialized characters and are independent in origin. Sections *Himalayicae* (Ulbr.) Juz., *Anemone*, *Omalocarpus* DC. and *Fuscopurpurea* Tarasevich et Chaudhary share common origin. Relatively primitive section *Himalayicae* and highly specialized section *Sylvia* are discussed in light of comparative study of above parameters.

INTRODUCTION

The genus *Anemone* L. is chiefly concentrated to the northern hemisphere and includes about 120 species. A number of species belonging to this genus are polymorphic and differentiated into many intraspecific taxa (1, 2, 8).

The system of the genus *Anemone* was first proposed by De Candolle(5). Monographic works carried out later by a number of systematists added new sections and series to this genus (3, 4, 6, 7, 11, 13, 16, 17, 18, 19). However, no attempt is made to study the evolution of infrageneric taxa of the genus *Anemone*. In this paper, probable direction of evolution and phylogenetic relationship of infrageneric taxa based on comparative study of rhizome, leaf, inflorescence, nutlet and pollen grains of Nepalese representatives and the adjacent regions of the Himalayas are discussed.

MATERIALS AND METHODS

Study and collection of material were done during field expedition in central Nepal in 1985. In addition, over 5000 specimens of herbaria meterial deposited in the herbarium section of BM, BSD, CAL, DD, DE, E, K, KATH, LE, P, S were examined. A comparative morphological study of rhizome, leaf, inflorescence, nutlet and pollen grains was done. Study of nutlet and pollen grains was done with the help of light microscope (MBR-1), scanning electron microscope (JSM-35C, Jeol) and transmission electron microscope (Tesla-BS-500) at Komarov Botanical Institute, Leningrad, USSR.

RESULTS AND DISCUSSION

A synopsis of the genus *Anemone* in Nepal and adjacent regions of the Himalayas was given by Chaudhary (3) who disposed 14 species, 19 subspecies and 3 varieties into following 6 sections and 4 series: Sect. 1 *Himalayicae* (Ulbr.) Juz. (4 sp., 9 subsp., 3 var.); Sect. 2. *Omalocarpus* DC., Ser. 1. *Involucratae* Ulbr. (3 sp., 4 subsp.), Ser. 2. *Involucellatae* Ulbr. (1 sp.); Sect. 3. *Fuscopurpurea* Tarasevich et Chaudhary (1 sp.); Sect. 4 *Anemone*, Ser. 1 *Rupicolae* Tamura ex Chaudhary et Trifonova (2 sp., 4 subsp), Ser. 2 *Anemonospermos* (DC.) Ulbr. (1 sp.); Sect. 5 *Rivularidium* Jancz. (1 sp. 2 subsp.); Sect. 6. *Sylvia* Spach. (1 sp.).

The study is based on regional representatives of the genus, however, the characters possessed by rhizome, leaf, inflorescence, nutlet and pollen grains reflect relative level of specialization of the sections and their phylogenetic relationship.

Rhizome in Sect. *Sylvia (A. griffithii)* is characterized by highly reduced internodes among the taxa studied and probably represent high level of organization in this group.

Leaf form in the species of the genus is polymorphic (Fig. 1). It is typically 3-lobed or 3-partite with lobes sessile (sect. *Anemone, Rivularidium, Fuscopurpurea, Omalocarpus* and few species of Sect. *Himalayicae*). Leaf with petiolate lobes (Sect. *Sylvia*) or undivided in Sect. *Himalayicae (pro parte)* are considered secondary in origin.

Scheme of inflorescence is given in Fig. 2. Cyme type of inflorescence which is characteristic in *Ranunculaceae* is present in the species belonging to Sect. *Rivularidium*, Ser., *Involucellatae* (Sect. *Omalocarpus*) and Ser. *Anemonospermos* (Sect. *Anemone*). Species belonging to Ser. *Involucratae* (Sect. *Omalocarpus*) and *A. fuscopurpurea* (Sect. *Fuscopurpurea*) possess um-

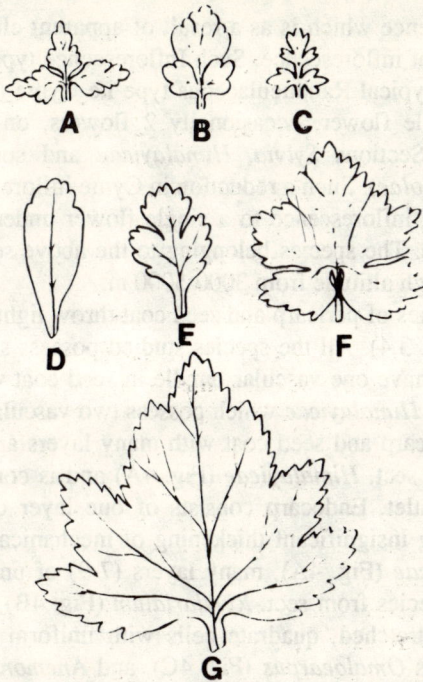

Fig. 1: Leaf form of some species of the genus *Anemone*: A—Sect. *Sylvia* (*A. griffithii*); B, C, D, E—Sect. *Himalayicae* (B—*A. geum* subsp. *ovalifolia*, C—*A. rupestris*, D—*A. trullifolia* subsp. *holophylla*, E—*A. trullifolia* subsp. *trullifolia*); F—Sect. *Omalocarpus* (*A. polyanthes*); G—Sect. *Anemone* (*A. vitifolia*).

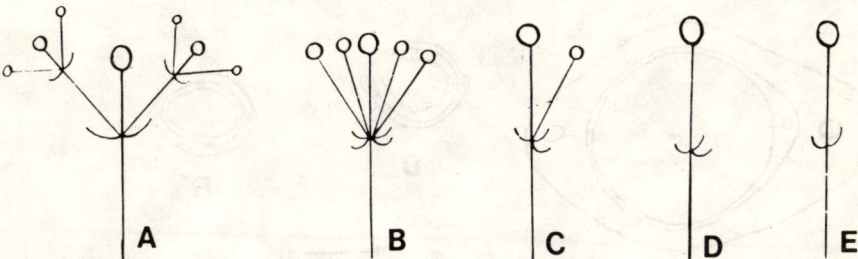

Fig. 2: Scheme of inflorescence in the genus *Anemone*: A=Sect. *Rivularidium*, Sect. *Omalocarpus ser. Involucellatate*. Sect. *Anemone* Ser. *Anemonospermos*; B-Sect. *Fuscopurpurea*, Sect. *Omalocarpus* Ser. *Involucratae*; C, D-Sect. *Himalayicae*, Sect. *Anemone* Ser. *Rupicolae*; E- Sect. *Sylvia*.

bel type of inflorescence which is as a result of apparent elimination of inter-nodes of cyme type of inflorescence. Such Inflorescence type is secondary and is derived from the typical Ranunculaceous type i.e Cyme (14). Inflorescence consisting of a single flower, occasionally 2 flowers, on the main axis is represented in the Sections *Sylvia*, *Himalayicae* and some taxa in Sect. *Anemone* (Ser. *Rupicolae*). Such a reduction in Cyme inflorescence might have brought reduction of inflorescence to a single flower under the stress condi-tions of high altitude. The species belonging to the above sections are chiefly found relatively at high altitude from 3000-5000 m.

Anatomical studies of pericarp and seed coat throw light on the evolution-ary relationship (Fig. 3,4). All the species studied possess specialized charac-ters of nutlet which have one vascular bundle in seed coat with the exception of species from sect. *Himalayicae* which possess two vascular bundles (Fig. 3) (9). In addition, pericarp and seed coat with many layers are characteristic to species belonging to sect. *Himalayicae* (Fig. 4A) and is considered relatively primitive type of nutlet. Endocarp consists of one layer of uniform, small, quadrate cells having insignificant thickening of mechanical tissue in species from sect. *Himalayicae* (Fig. 4A), many layers (7-9) of uniformly thick me-chanical tissue in species from sect. *Rivularidium* (Fig. 4B), 1 layer (rarely 2-layers) of radially stretched, quadrate cells with uniform thickening in the species from sections *Omalocarpus* (Fig. 4C), and *Anemone* (Fig. 4D) and 1 layer of radially stretched thickening in the sect. *Sylvia* (Fig. 4E).

In the species studied, pollen grains are 3-colpate and pantolpate (Fig. 5

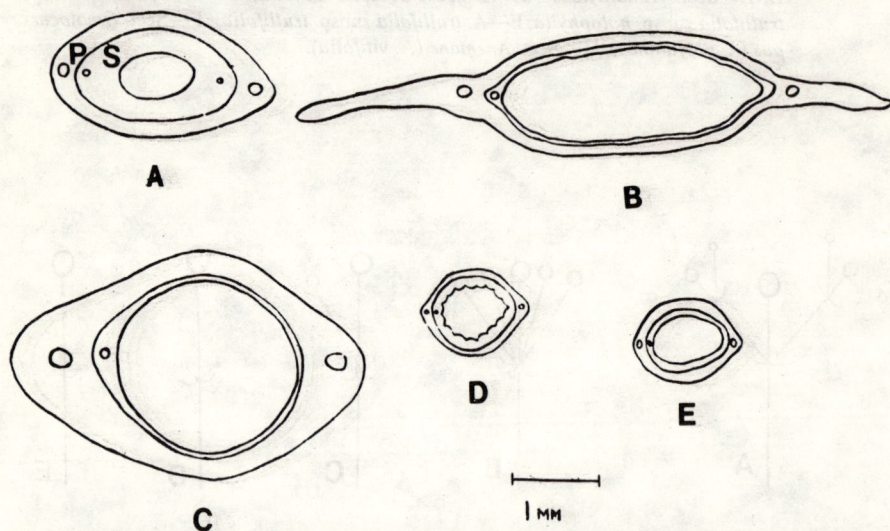

Fig. 3.   Scheme of cross section of nutlet of some species of the genus *Anemone*:- A–Sect. *Himalayicae* (*A. obtusiloba*); B–Sect. *Omalocarpus* (*A. narcissiflora*); C–Sect. *Rivularidium* (*A. rivularis*); D–Sect. *Anamone* (*A. vitifolia*); E–Sect. *Sylvia* (*A. nemo-rosa*) (P–Pericarp, S–Seed coat).

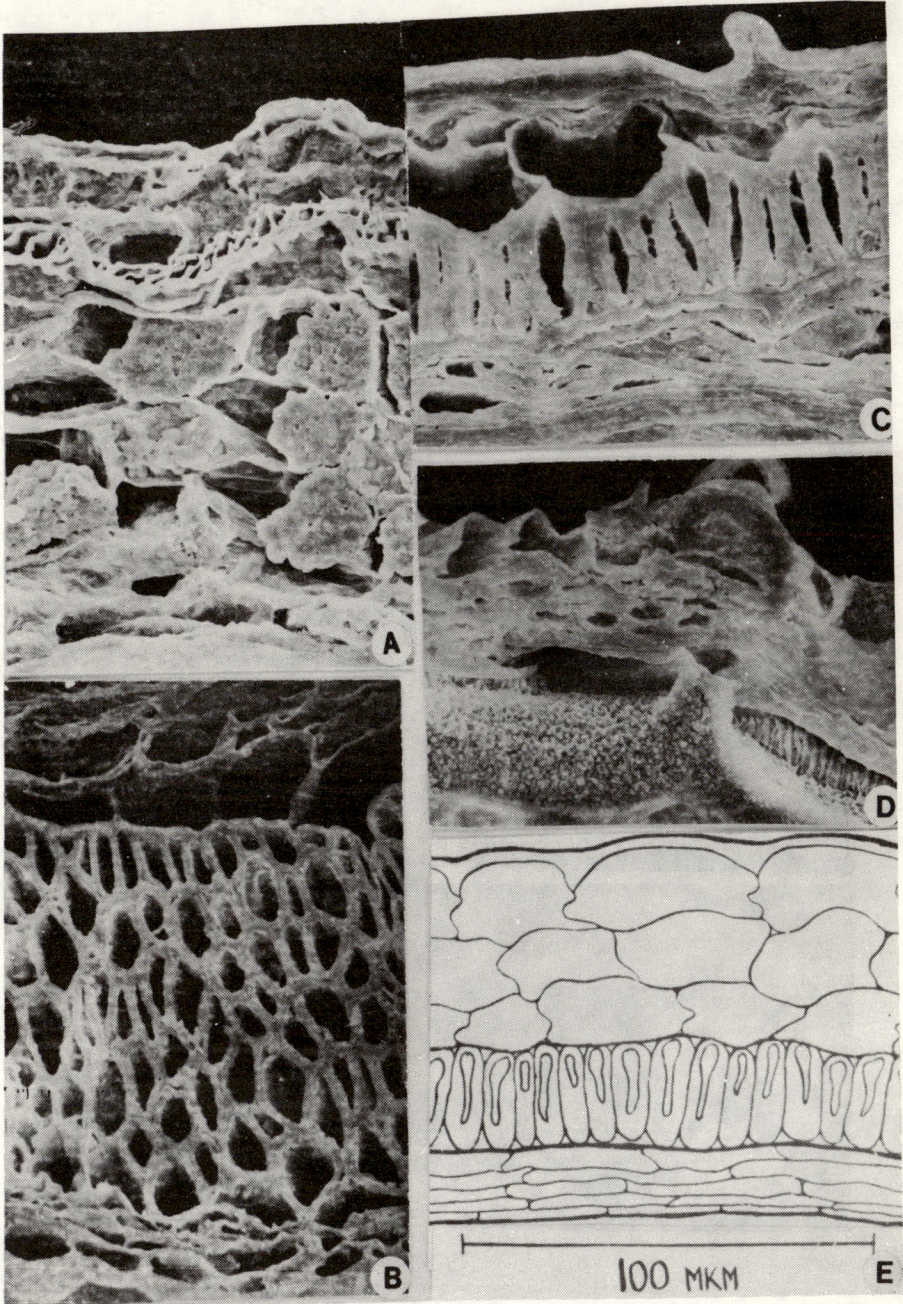

Fig. 4: Anatomical structure of nutlet (Pericarp and Seed coat) of the genus *Anemone*: A—Sect. *Himalayicae* (*A obtusiloba,* SEM × 320); B-Sect. *Rivularidium* (A, rivularis, SEM x 320); C—Sect. *Omalocarpus* (*A. demissa,* SEM × 660); D—Sect. *Anemone* (*A. vitifolia,* SEM × 660); E—Sect. *Sylvia* (*A. nemorosa*-Scheme).

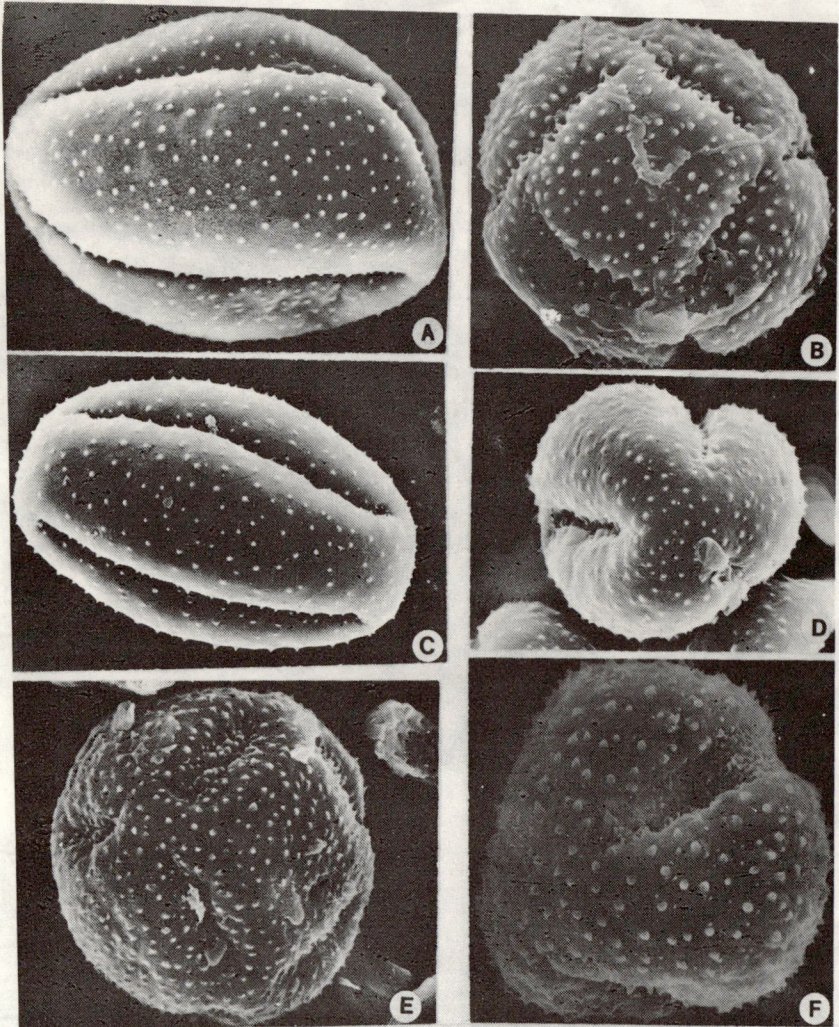

Fig. 5: Structure of Pollen grains of genus *Anemone* under SEM (all magnification × 2200); A—Sect. *Himalayicae* (*A. obtusiloba*); B—Sect. *Sylvia* (*A. griffithii*); C—Sect. *Omalocarpus* (*A. polyanthes*); D—Sect. *Anemone* (*A. vitifolia*); E—Sect. *Rivularidium* (*A. rivularis*); F—Sect. *Fuscopurpurea* (*A. fuscopurpurea*).

A-E). Evolutionary direction of pollen grains in the genus show that 3-colpate pollen grains which is characteristic to majority of species studied comprise relatively primitive character in comparison to pantocolpate pollen grain in sections *Sylvia (A. griffithii)* (Fig. 5B), and *Rivularidium (A. rivularis)* (Fig. 5E) (12, 17). Sexine structure of the pollen grains is tectate perforate with echinate sculpture in all the species studied (Fig. 5A-E) except *A. fuscopurpurea* which possess tuberculato echinate sculpture (Fig. 5F).

Thus, considering the characters of rhizome, leaf, inflorescence, nutlet and pollen grains, it has been seen that each section comprises together a mosaic combination of characters of quite different evolutionary levels. However, in totality, putative phylogenetic relationship and relative level of specialization of sections in the genus *Anemone* are given in Fig. 6.

In my opinion, relatively less specialized characters of organs are possessed by sect. *Himalayicae* (rhizome with long internodes, 3-lobed/3-partite leaf, less specialized characters of pericarp and seed coat, 3-colpate pollen grains, etc.) and Sect. *Sylvia* comprises more specialized characters among the taxa studied on account of highly reduced internodes of rhizome, inflorescence of a single flower, 1 layer of endocarp having unequal thickenings, polycolpate pollen grains, etc. while the secions *Rivularidium, Fuscopurpurea, Omalocarpus* and *Anemone* occupy intermediate positions. Section *Rivularidium (A. rivularis)*, besides containing many layers of endocarp, is characterized by unique fourth pair of chromosome with short arms (10) and dorsoventrally oriented vascular bundles in the petiole (15). Sections *Himalayicae, Anemone, Omalocarpus* and *Fuscopurpurea* probably have originated from a common ancestor while sections *Sylvia* and *Rivularidium* constitute a separate lines of phylogeny, independently and distantly from each other.

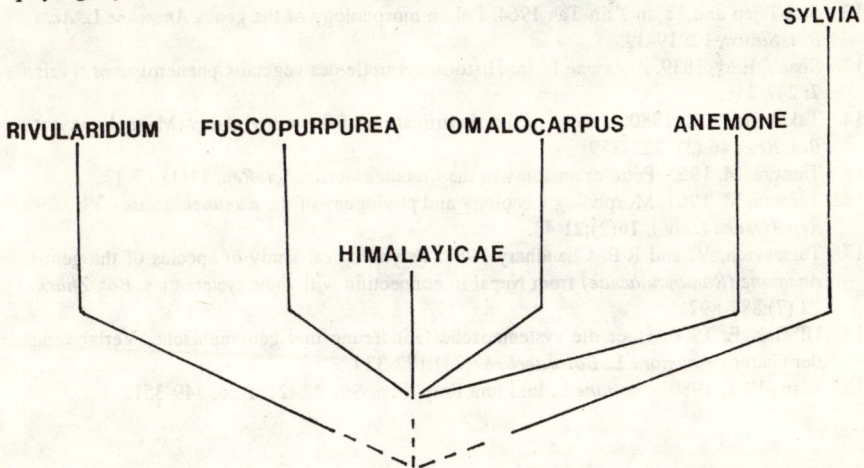

Fig. 6: Scheme of putative phylogenetic relationship of sections in *Anemone*.

Acknowledgements

I express my sincere thanks to Prof. Armen L. Takhtagan for supervision of my Ph.D. dissertation. The curators of the above named herbaria are thankfully acknowledged for the loan of the plant materials.

REFERENCES

1. Chaudhary, R.P. 1986. Intraspecific variation in *Anemone rupicola* Cambess L., and allied species from the Himalayas. Proceedings of the Ist Youth Botanical Conference, Leningrad (USSR); 104-107 (In Russian).
2. Chaudhary, R.P. 1987. Taxonomy of *Anemone rupicola (Ranunculaceae)* and a new species from the Himalayas. *Bot. Zhurn.* 72 (6): 820-827.
3. Chaudhary, R.P. 1988. A synopsis of the genus *Anemone (Ranunculaceae)* in Nepal and adjacent regions of the Himalaya. *Bot. Zhurn.* 73 (8): 1188-1202.
4. Chaudhary, R.P. and V.I. Trifonova, 1988. Morphology of fruit and comparative anatomy of pericarp and seed coat in the Nepalese species of the genus *Anemone (Ranunculaceae). Bot.Zhurn.* 73 (6): 803-817.
5. De Candolle, A.P. 1818, Regni Vegetabilis Systema Naturale. I. Paris.
6. Hooker, J.D. and T. Thomson, 1855. *Ranunculaceae*, In: Flora Indica, 1: 1-25.
7. Janczewski, E. 1890. Éetudes comparées sur le genre *Anemone. I. Fruit, II. Germination. Bul. Intern. Acad. Sci. Crakovie* (Sci. Nat.): 298-303.
8. Lauener, L.A. 1960. Notes on *Anemone obtusiloba* and its allies. *Not. Roy. Bot. Gard. Edinb.*, 23(2): 179-201.
9. Lonay H. 1901. Contribution a l' anatomie Renonculees. *Arch. Soc. Roy. Sci. Liege*, 3: 1-190.
10. Moffet, A.A. 1932. Chromosome studies in *Anemone* I. A new type of chiasma behaviour. *Cytologia*, 4: 26-37.
11. Prantl, K. 1888. Beitrage zur morphologie und systematik Pflanzengeschichte und Pflanzengeographie, Leipzig, 9: 225-273.
12. Si I-Tsjen and Tsjan Tsin-Tan 1964. Pollen morphology of the genus *Anemone* L. *Acta. Bot. Sinica*, 12: 19-39.
13. Spach, E.M. 1839. *Anemone* L. In: Histoire naturelle des végétaux phanerogames, Paris, 7: 242-256.
14. Takhtajan, A.L. 1980. Outline of the classification of flowering plants (Magnoliophyta). *Bot. Rev.*, 46 (3): 225-359.
15. Tamura, M. 1962. Petioler anatomy in the *Ranunculaceae. Sci. Rep.* 11(1):17-47.
16. Tamura, M. 1967. Morphology, ecology and phylogeny of the *Ranunculaceae* - VII. *Sci. Rep. (Osaka Univ.)*, 16(2):21-43.
17. Tarasevich, V. and R.P. Chaudhary 1987. Palynological study of species of the genus *Anemone (Ranunculáceae)* from Nepal in connection with their systematics. *Bot. Zhurn.* 72 (7):887-897.
18. Ulbrich, E. 1906. Über die systematische Gliederung und geographische, Verbreitung der Gattung *Anemone* L. *Bot. Jahrb.* 37(23):172-334.
19. Wang W.T. 1980. *Anemone* L. In: Flora Reip. Pop. Sin. 28 (2): 1-56, 349-351.

17

Response of Improved Wheat Cultivars of Nepal to a Change in Environment

Sabitri Shrestha

Central Department of Botany, Tribhuvan University, Kirtipur,
Kathmandu, Nepal

ABSTRACT

Eleven improved wheat cultivars collected from Kathmandu (Nepal) were cultivated in the different environmental conditions of Moscow (USSR) and Kathmandu and their responses in growth and development were observed. In the new
environment the wheat cultivars showed the following changes:
1. The vegetation period became too short.
2. The plant became dwarfed with a short spike length.
3. The spiklet numbers also reduced, which caused a reduction of grains in the
   spike.
4. The single grain weight changed insignificantly.
All these results are related to the differential effects of temperature and light
intensity in the two countries.

INTRODUCTION

In Nepal wheat holds third position after rice and maize while in Moscow
(USSR) it is first in the ranking of crops. It is generally cultivated in the winter
season in Nepal despite a spring-type mode of life cycle. Sowing is done in
October, November and December and harvesting in May. In Moscow, sowing
is generally done in early May and harvesting in August. Hence the times of
sowing and harvesting in the two countries contrast markedly as do temperature, photoperiod and insolation [1].

The main objective of this research was to establish the effect of environment (especially temperature and photoperiod) on the growth and development of the improved wheat cultivars of Nepal whose origins are quite different (Table 1).

Table 1: Characters of tested wheat cultivars

Cultivars	Origin	Year of recomm-endation	Zones of Nepal	Days of maturty	1000 grain wt. (gm)	Grain colour
1. Lerma-52	Colombia	1959/60	Hills	176	41	White
2. Lermarajo-64	Mexico	1965/66	Hills	168	40	Red
3. RR-21	India	1970/71	Whole Nepal	116	49	White
4. UP-262	India	1978/79	Terai	122	46	White
5. Siddhartha	India	1985/86	Terai	118	42	White
6. Vinayak	India	1988/89	Terai	120	44	White
7. NL-297	-	1985/86	Terai	-	-	-
8. Bhaskar	Mexico	1983/84	Terai	-	-	-
9. Lumbeni	India	1981/82	Whole Terai	120	47	White
10. Tribeni	India	1982/83	Terai	120	42	White
11. NL-30	Nepal	1983/84	Western Part of Nepal	120	47	White

Source: Agriculture Diary (1943-44)

MATERIALS AND METHODS

Wheat cultivars of different origin, viz., Lerma-52, Lermarajo-64, RR-21, UP-262, Siddhartha, Vinayak, NL-297, Bhaskar, Lumbini, Tribeni and NL-30 were provided by the Khumaltar agricultural farm, where the local experiments were carried out in 1985. The climate is warm and humid in the summer and dry cold in the winter season.

The foreign experiments were conducted in the Biological Station of Zhveningorod of Moscow State University, situated near the Moscow (Moskva) River. The climate here is continental. Winter is very cold with snowfalls (150 to 160 days from late October to March). From May to September the climate is warm with an average temperature of 14°C.

In Kathmandu, sowing was done in November 1985 in a randomised complete block design. There were three replications. Individual plots were 2.5 m × 1.0 m and contained 5 rows with crop geometry 25 cm × 10 cm. The seeds were sown at the rate of 2 seeds/hill with the recommended dose of fertilisation (NPK-100, 60, 40). Uniformity of plant stand was maintained by thinning.

The same experiment was repeated in Moscow except that sowing was done in April 1986, the usual sowing period in this area.

Special attention was given to phenological characters. Some developmental periods were also noted, i.e. emergence to heading and heading to maturation. The developmental periods of all the cultivars in Moscow were compared with those in Kathmandu (Fig. 1). The yield components were also studied. Observations and results were compiled on 50 plants from the middle three rows in each of the three replicates and were selected randomly. The parameters of biometric measurements were plant height, spike length, number of spiklets, grains per spike, 1000 grain weight and single grain weight. The results of biometric measurements of different agronomic traits were tabulated in mean value (Table 2).

Table 2: Effect of environment of yield component of wheat cultivars in Kathmandu and Moscow (mean value)

Cultivars	Place of cultivar	Plant height (cm)	Spike length (cm)	Spiklet number	No. of grains spike	Single grains wt (mg)	Grain weight spike(gm)
Lerma-52	*Kath.	75.3	9.0	16.0	41.0	40.8	-
	Mos.	55.7	6.7	10.4	14.4	40.2	0.58
Lermarajo 64	Mos.	50.6	8.2	11.5	19.9	45.7	0.86
	Kath.	64.8	8.8	15.0	25.0	53.4	-
RR-21	Mos.	52.1	7.8	8.5	10.0	42.6	0.42
UP-262	Kath	60.8	8.5	19.0	29.0	48.5	-
	Mos.	39.0	6.7	13.5	21.1	47.0	1.00
Siddhartha	Kath.	47.9	7.0	18.0	26.0	36.5	-
	Mos.	44.1	6.6	12.8	23.8	35.0	0.84
Vinayak	Kath.	41.8	7.2	16.0	27.0	42.2	-
	Mos.	47.0	6.6	12.0	18.5	38.0	0.72
NL-297	Mos.	50.1	7.1	11.0	18.7	45.0	0.85
Bhaskar	Kath	51.4	7.7	17.0	26.0	38.0	-
	Mos.	51.1	7.0	13.2	23.9	36.3	0.87
Lumbini	Kath	57.3	8.5	18.0	35.0	44.3	-
	Mos.	43.8	7.8	11.2	17.6	53.7	0.91
NL-30	Kath.	60.3	8.0	17.0	34.0	41.8	-
Tribeni	Mos.	62.6	6.5	11.5	23.2	33.5	0.78

*Kath.: Kathmandu (Khumaltar agriculture farm)
Mos.: Zhveningorod Biological Station of Moscow

RESULTS

On comparing the developmental stages of wheat cultivars, it was found that their life cycles or vegetative period were shortened in Moscow compared to Kathmandu. During the course of development, the winter-sown wheat cultivars in Nepal took a longer time from sowing to heading than in Moscow. In all the tested wheat cultivars this time fluctuated from 78 to 130 days. In

Special attention was given to physiological characters. Some developmental periods were also noted: the emergence to heading, and heading to maturation. The developmental periods of all the cultivars in Moscow were compared with those in Kathmandu (Fig. 1). The yield components were also studied. Observations and results were counted on 50 plants from the main tiller, three rows in each of the three replicates and were statistically analyzed. Characters of biometric measurements were plant height, number of tillers per plant, spike length, number of spikelets per spike, number of grains per spike. The results of biometric measurements of different agronomic traits were tabulated in mean value (Table 2).

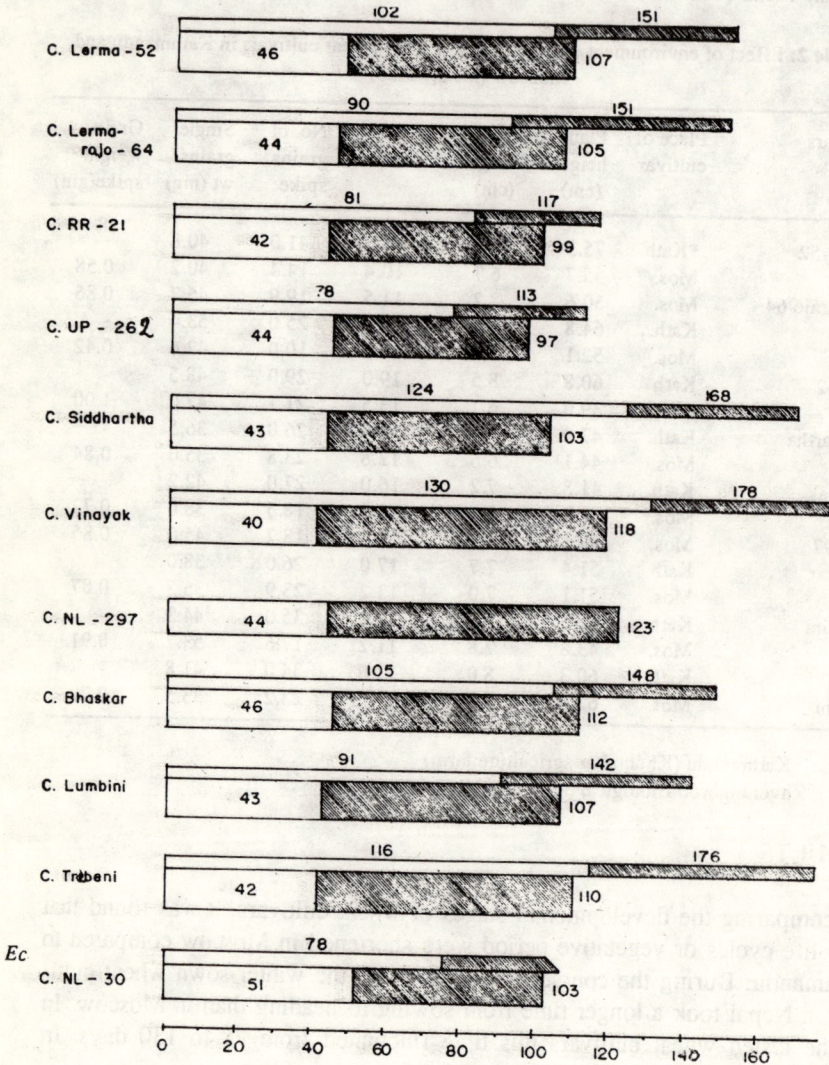

Period from sowing to heading

Period from heading to maturation

Fig. 1: Comparison of the rate of development of tested cultivars in Nepal and Moscow.

160   *Sabitri Shrestha*

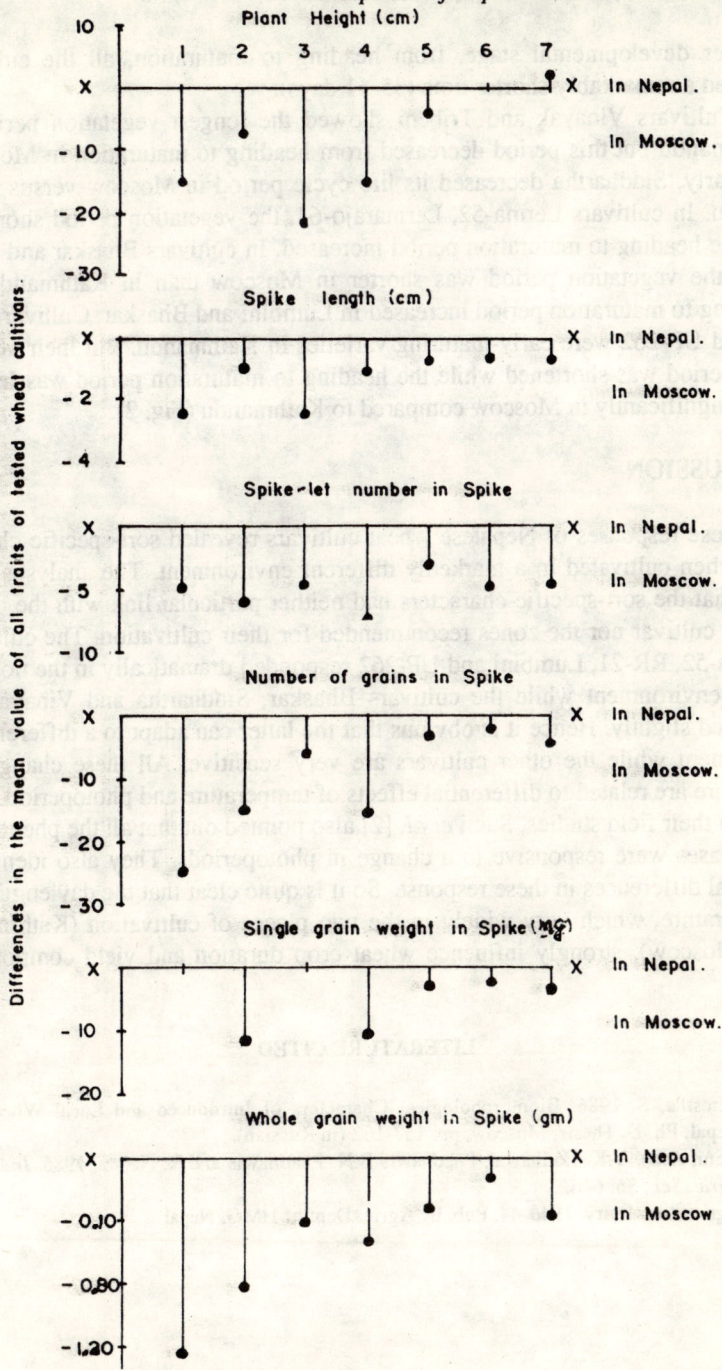

Fig. 2: Comparative study of different traits of matured plants experimentally tested in Nepal and Moscow.

another developmental stage, from heading to maturation, all the cultivars showed a remarkably shorter time (35-51 days).

Cultivars Vinayak and Tribeni showed the longest vegetation period in Kathmandu but this period decreased from heading to maturation in Moscow. Similarly, Siddhartha decreased its life cycle period in Moscow versus Kathmandu. In cultivars Lerma-52, Lermarajo-64, the vegetation period shortened but the heading to maturation period increased. In cultivars Bhaskar and Lumbini, the vegetation period was shorter in Moscow than in Kathmandu but heading to maturation period increased in Lumbini and Bhaskar. Cultivars RR-21 and UP-262 were early-maturing varieties in Kathmandu. But their vegetation period was shortened while the heading to maturation period was lengthened significantly in Moscow compared to Kathmandu (Fig. 2).

DISCUSSION

All these responses of Nepalese wheat cultivars revealed sort-specific characters when cultivated in a markedly different environment. The analyses indicate that the sort-specific characters had neither particular link with the origin of the cultivar nor the zones recommended for their cultivation. The cultivars Lerma-52, RR-21, Lumbini and UP-262 responded dramatically in the non-optimal environment while the cultivars Bhaskar, Siddhartha and Vinayak responded slightly. Hence it is obvious that the latter can adapt to a different environment while the other cultivars are very sensitive. All these changes in structure are related to differential effects of temperature and photoperiod.

In their field studies, Saini *et al.* [2] also pointed out that all the phenological phases were responsive to a change in photoperiods. They also identified varietal differences in these response. So it is quite clear that the daylength and temperature, which vary widely in the two places of cultivation (Kathmandu and Moscow), strongly influence wheat crop duration and yield components [2].

LITERATURE CITED

1. Shrestha, S. 1986. Biomorphological Characters of Introduced and Local Wheats of Nepal, Ph. D. Thesis, Moscow, pp. 157-162 (in Russian).
2. Saini, A.D., V.K. Dadhawal, Phadnawis B.N. Phadnawis and R. Nanda, 1986. *Indian J. Agric. Sci.,* **56**: 646.
3. Agriculture Diary. 1943-44. Pub. by Agric. Dept of HMG, Nepal.

18

Evaluation of Wheat Genotypes Adapted to Nepal

R.C. Sharma and N.K. Chaudhary

Institute of Agriculture and Animal Science, Rampur, Chitwan, Nepal

ABSTRACT

A set of 15 spring wheat (*Triticum aestivum* L.) genotypes (11 released cultivars and 4 advanced breeding lines) were studied to examine the range of variation for several vegetative and grain yield traits, and to determine the traits with profound effects on grain yield. The 15 genotypes were evaluated in a replicated field test at Rampur, Chitwan, Nepal in 1988. The results indicated a range of variation for vegetative traits, grain yield components, and grain yield. The newer genotypes usually showed increased kernel weight which had a significant positive correlation with grain yield. Biomass yield appeared to have the most dominant effect on grain yield compared to other agromatic traits.

INTRODUCTION

In the past 25 years, wheat yield improvemnt in Nepal has been achieved through selection for yield per se. Because of the slow response to selection for yield, there is a need for ascertaining some alternative indirect selection criteria to increase grain yield.

A previous study revealed that the leading wheat cultivars in Nepal exhibited variation for maturity [1]; however, the relationship between maturity and grain yield needs to be determined.

Grain-filling duration is a critical trait affecting wheat yields in Nepal because planting of wheat after the harvest of the late-maturing paddy (*Oryza sativa* L.) cultivars reduces both the days in vegetative growth period and grain-filling period of wheat.

Biomass yield of a wheat genotype represents its total photosynthetic capacity and, at a given level of harvest index, a higher biomass yield will result in a higher grain yield. The present-day wheat cultivars of Nepal show differences in their biomass yield [1].

The Harvest Index has been shown to have a strong positive correlation with grain yield [2, 3]. As yet, there is no information available on HI of the wheat cultivars in Nepal. Also, the level of genotype × environment interaction for HI, compared to grain yield, needs to be investigated under the diverse environmental conditions of Nepal.

Three grain yield components in cereals include the number of spikes per unit area, number of kernels per spike, and average kernel weight. There is a developmental sequence for these three components and their association with grain yield. Moreover, the presence of component compensation is clearly established [4].

This study was conducted with the following objectives:
1. To determine the range of variation for agronomic traits in a series of wheat genotypes adapted to Nepal.
2. To investigate the correlations among agronomic traits and to identify the one or more traits with pronounced effect on grain yield.

MATERIALS AND METHODS

A set of 15 spring wheat genotypes (pure-line) was chosen for this study. Included were 11 commercial cultivars, released in Nepal over the past 25 years, and four advanced breeding lines. The name and the year of release of these wheat genotypes are given in Table 1.

The genotypes were grown at the agronomy research farm of the Institute of Agricultural and Animal Science, Rampur, Nepal in the 1988 wheat-growing season (Table 2). The plot size was 1 × 5 m and planted as four 5-m long rows at 0.25-m row spacing. The plots were seeded on 4 December, 1987 at the seeding rate of 100 kg/ha. Fertilisers, at the rates of 80, 40, and 20 kg/ha respectively of N, P, and K, were applied prior to seedling. A top dressing with urea at the rate of 20 kg/ha of N was done at the joining stage. Because of uneven weed development in the plots, one hoeing was done at the joining stage.

The number of seedlings in the two centre rows of each plot was counted at 17 days after seeding to determine the initial seedling number per unit area (LSN). The maximum number of tillers (MTN) in each plot was also determined from periodic tiller counts in the two centre rows. Twenty primary tillers were tagged and the number of leaves (LFT), above-ground nodes (NDT) and peduncle length (PDL) were recorded on 10 primary tillers ran-

Table 1: Name and the year of release of the 15 spring wheat genotypes included in this study

Genotype	Year of release
Lerma 52	1960
Lermarajo 64	1965
RR 21	1968
NL 30	1975
HD 1982	1975
UP 262	1978
Lumbini	1981
Triveni	1982
Vinayak	1983
Siddhartha	1983
Vaskar	1983
HUW 251	Advanced line
BL 1022	Advanced line
BL 1949	Advanced line
NL 549	Advanced line

Table 2: List of the characters and their representative symbols in the study of 15 spring wheat genotypes

Character	Symbol
Initial seedling number (m^{-2})	SLN
Maximum tiller number (m^{-2})	MTN
Number of nodes per tiller	NDT
Number of leaves per tiller	LFT
Peduncle length (cm)	PDL
Plant height (cm)	PHT
Length of vegetative growth period (days)	VGP
Length of grain-filling period (days)	GFP
Biomass yield (kg ha^{-1})	BMY
Grain yield (kg ha^{-1})	GRY
Harvest index (%)	HI
Number of spikes (m^{-2})	SPN
Spike length (cm)	SPL
Number of spikelets per spike	SPLPS
Number of kernels per spike	KPS
Thousand kernel weight	THKWT

domly chosen from the 20 tagged tillers. At maturity, plant height (PHT) was recorded in each plot. The length of the vegetative growth period (VGP) was recorded as sowing to heading period. Duration of grain-filling period (GFP) was recorded as the number of days from heading to physiological maturity (when the plant loses its green colour). The number of seed-bearing spikes was counted from the two centre rows of each plot. At maturity, the two centre rows of each plot were harvested close to the ground and weighed individually to record the biomass yield (BMY), grain yield (GRY) and per cent harvest index.

Variance analyses were conducted to determine genotypic differences for the 16 characters and simple correlation coefficients were calculated among these characters.

RESULTS AND DISCUSSION

Significant genotypic differences were established for all 16 traits included in this study (Table 3). This suggests that an array of variations is available for all these traits among the leading wheat genotypes grown in Nepal.

Mean values for individual traits are presented in Table 4. The SLN showed a range between 135 and 222 m^{-2} which could be attributed to the differences in seed size, germination and vigour. The cultivar 'RR 21' had the lowest SLN while 'Vaskar' had the highest SLN.

The MTN produced by wheat genotypes ranged from 223 to 379 M^{-2}. The cultivar 'Lerma 52' had the highest MTN compared to the lowest for 'BL 1022'. The genotypes with the highest and the lowest SLN did not possess the maximum and minimum values for tiller number, which indicates that the genotypes differed in their tillering capacity.

The average number of NDT and LFT on the main culm ranged, respectively, from 4.3 to 5.6, and 5.9 to 8.4, which can be considered a relatively low range of variation. Most of these genotypes were semi-dwarf except the tall Lerma 52; the narrow ranges for LFT and NDT were as expected. The genotype 'NL 549' showed the lowest values for both these traits.

The PDL and PHT of the 15 genotypes varied from 35.0 to 45.1 cm and 76 to 114 cm respectively. The tallest genotype, Lerma 52, did not have the longest PDL, while one of the shortest genotypes, 'Vinayak', had the shortest peduncle. The PDL of the wheat genotypes accounted for as much as 40% of the entire plant length. The peduncle, being right next to the spike, might contribute to the grain development by producing photosynthates, particularly under the early senescence of the flag leaf that usually occurs under high temperatures and/or foliage disease development.

The VGP range between 63 to 77 days with RR 21 having the shortest VGP and Lerma 52 having the longer VGP. A shorter VGP followed by a longer GFP could be advantageous for grain yield stability of wheat under shorter spring duration on the plains of Nepal. Of all the semi-dwarf wheats grown in the Indian subcontinent, the cultivar RR 21 is considered the one with the widest adaptability.

The average GFP varied from 42 to 52.8 days. The genotype Lerma 52, with the longest duration of VGP, showed the shortest GFP. This range of

Table 3: Analysis of variance for 16 traits in 15 spring wheat genotypes

Source	df	SLN⁺	MTN	NDN	LFN	PDL	PHT	VGP	GFP	BMY	GRY	HI	SPN	SPL	SPLPS	KPS	THKWT
										Mean squares							
Replication	03	ns⁺⁺	ns	ns	ns	ns	ns	ns	**	**	**	ns	ns	ns	ns	**	**
Genotypes	14	**	**	**	**	**	**	**	**	**	**	**	**	**	**	**	**
Error	42	131	1149	0.11	0.31	4	25	1.1	2.3	260242	59184	5.2	773	0.12	0.62	8.2	4

+ For explanation of the symbols, see Table 2.
++ Non-significant at the 0.05 probability level.
** Significant at the 0.01 probability level.

Table 4: Mean values for 16 characters in 15 spring wheat genotypes

Genotypes	SLN+	MTN	NDT	LFT	PDL	PHT	VGP	GFP	BMY	GRY	HI	SPN	SPL	SPLPS	KPS	THKWT
Lerma 52	219	379	5.6	8.4	41.0	114	76.8	42.0	5235	1590	30.4	248	8.8	14.0	34.2	33.7
Lermarajo 64	210	320	4.8	6.8	40.0	95	64.8	49.3	5895	2310	39.2	272	9.5	13.9	36.9	38.3
RR 21	135	240	5.0	6.8	41.8	96	63.0	50.8	5970	2375	39.8	208	9.9	13.8	34.3	46.1
NL 30	168	298	5.0	7.8	39.9	96	71.0	46.5	5595	2180	39.0	242	9.5	15.1	36.9	38.9
HD 1982	190	288	4.6	6.9	36.5	86	64.8	48.8	5925	2405	40.6	258	9.5	15.9	41.3	41.0
UP 262	201	319	5.0	7.4	37.0	94	69.5	44.3	7190	2815	39.2	252	9.1	16.7	38.0	46.5
Lumbini	198	335	4.5	6.2	36.2	85	64.5	49.8	6215	2480	39.9	276	8.7	15.6	42.3	39.3
Triveni	174	225	4.9	7.2	40.8	92	71.3	43.3	4860	2010	41.4	185	9.4	16.3	48.3	42.9
Vinayak	169	279	4.7	6.3	35.0	80	67.0	49.3	6230	2420	38.8	251	8.7	13.8	40.9	43.1
Siddhartha	165	246	4.6	6.1	35.4	76	65.0	52.8	5370	2305	42.9	229	8.6	15.6	39.4	42.3
Vaskar	222	320	4.6	6.0	37.6	80	71.3	43.3	6210	2555	41.1	249	9.0	16.1	45.0	37.3
BL 1049	211	260	5.0	6.3	37.0	94	67.5	43.8	6185	2410	39.0	241	8.8	15.7	34.7	46.9
HUW 251	177	223	4.7	6.4	43.7	96	68.5	45.0	6435	2635	40.9	222	9.3	16.4	39.1	49.0
NL 549	186	276	4.8	7.1	38.4	90	64.8	44.3	4370	1745	39.9	199	8.2	16.3	46.6	37.3
BL 1022	148	246	4.3	5.9	45.1	86	66.5	49.3	6835	2685	39.3	189	9.0	17.5	51.1	48.2
Average	185	283	4.8	6.9	39.1	91	68.0	46.8	5901	2328	39.4	236	9.1	15.5	40.6	42.1
LSD.05	17	51	0.5	0.8	3.0	7	1.6	2.3	774	369	3.4	42	0.5	1.2	4.3	3.0
C.V. (%)	6.2	11.9	8.9	9.3	5.1	5.5	1.5	3.2	8.6	10.5	5.8	11.8	3.8	5.1	7.1	4.8

+ For explanation of the symbols, see Table 2.

10-day difference could result in yield advantage if the spring temperatures do not rise abruptly.

The biomass yield of the 15 genotypes showed a wide range of variation between 4370 and 7190 kg/ha [3]. The biomass variation in these 15 genotypes suggests that selection for this trait can be done among the progenies produced by crossing diverse parents.

Grain yield ranged between 1590 and 2815 kg/ha. The oldest among the 15 genotypes, Lerma 52, showed the lowest grain yield, while UP 262 showed the highest grain yield.

The harvest index ranged from 30.4 to 42.9%. The range of HI was rather narrow if Lerma 52 is not considered. The HI values of these genotypes suggest that this trait can be improved only by including wheat parents with higher HI from outside sources.

The SPN ranged between 185 to 272 m^{-2}. These values are rather low because the optimum SPN for maximising GRY is considered to lie in the range of 350 to 400 for spring wheat. In part, this might be achieved by increasing the seeding rates.

The SPL showed rather a narrow range between 8.2 to 9.9 cm. There appears little scope for improving this character by including these wheat genotypes in the crossing scheme. So, inclusion of suitable exotic germplasm should be considered in the breeding populations.

The SPLPS ranged from 13.8 to 17.5 while KPS showed variations between 34 to 51. The genotypes with the fewest SPLPS did not have always the fewest KPS and vice versa. This suggests that the number of kernels per spikelet also differs among genotypes.

The thousand-kernel weight ranged from 33.7 to 49.0 g. The standard height old cultivar Lerma 52 had the lightest kernels. The THKWT appears to have improved over the years as the newer genotypes have heavier kernels except for an old release, RR·21.

The significant values for simple correlation coefficient between pairs of traits are presented in Table 5.

The findings of this study indicate that a good range of variation for most agronomic characters occurs among the wheat genotypes adapted to Nepal. This envisages that good improvements can be accomplished by crossing the selected parents. Initial seedling number appears to have a bearing on the final spike count, and hence seedling rates need to be studied for any adjustment based on kernel weight. Neither VGP nor GFP appeared to have a significant bearing GRY. There appeared to be only a limited compensation among GRY components and only THKWT showed positive correlation with grain yield. However, BMY showed the most pronounced influence on GRY.

Table 5: Simple correlation coefficients⁺ among the pairs of characters in 15 wheat genotypes

Traits	MTN++	NDT	LFT	PDL	PHT	VGP	GFP	BMY	GRY	HI	SPN	SPL	SPLP	KPS	THKWT
SLN	0.70	0.31	0.20	-0.36	0.23	0.44	-0.58	0.00	-0.15	-0.36	0.62	-0.29	-0.03	-0.20	-0.52
MTN		0.40	0.45	-0.30	0.34	0.41	-0.25	0.04	-0.25	-0.66	0.72	-0.20	-0.34	-0.30	-0.73
NDT			0.84	0.07	0.84	0.61	-0.54	-0.32	-0.60	-0.74	0.08	0.11	-0.49	-0.65	-0.33
LFT				0.14	0.79	0.58	-0.48	-0.40	-0.63	-0.66	0.05	0.19	-0.30	-0.39	-0.46
PDL					0.50	0.19	-0.16	0.06	-0.03	-0.20	-0.57	0.42	0.19	0.17	0.29
PHT						0.54	-0.53	-0.15	-0.47	-0.77	-0.01	0.27	-0.30	-0.51	-0.19
VGP							-0.76	-0.10	-0.35	-0.61	-0.01	-0.04	0.00	-0.10	-0.33
GFP								0.17	0.33	0.41	0.01	0.14	-0.24	-0.03	0.21
BMY									0.90	0.08	0.12	0.24	0.21	-0.11	0.59
GRY										0.51	0.32	0.26	0.37	0.08	0.69
HI											0.19	0.12	0.43	0.40	0.42
SPN												-0.22	-0.42	-0.51	-0.39
SPL													-0.22	-0.28	0.26
SPLPS														0.67	0.39
KPS															0.06

+ Table value at P (0.05) is 0.50 and at P (0.01) is 0.62.

++For explanation of the symbols, see Table 2.

LITERATURE CITED

1. Regmi, K.R. 1986. Agronomic investigations: Plant nutrient aspect of wheat. In: 13th Winter Crop Workshop: Wheat Reports. National Wheat Development Programme, Bhairahawa, Nepal.
2. Singh, I.D., and N.C. Stoskopf. 1971. Harvest index in cereals. *Agronomy J.*, **63**:226.
3. Kulshrestha, V.P. and H.K. Jain. 1982. Eighty years of wheat breeding in India: past selection pressures and future prospects. *Z. Pflanzenzeucht.*, **89**:19-30.
4. Grafius, J.E., R.L. Thomas and J. Barnard. 1976. Effect of parental component complementation on yield and components of yield in barley. Crop Sci. **16**:673-677.

19

Hormonal Role in the Sterility of Rice
(*Oryza sativa* L.)

Santa B. Gurung

Institute of Agricultural and Animal Science,
Tribhuvan University, Rampur, Chitwan, Nepal

ABSTRACT

Two approved varieties of rice (*Oryza sativa* L.) viz., Jaya and Palman, recommended for cultivation in the Punjab were studied in the field and laboratory at Punjab Agric. Univ., Ludhiana. Indoleacetic acid (5 ppm), gibberellic acid (25 ppm), kinetin (10 ppm) and Ethrel (25 ppm) were sprayed at the panicle emergence and flowering stages of both varieties. Samples for determination of the endogenous levels of hormones were taken after 5 days of the first spraying and after 7 days of the second spraying. Grain yield contributing parameters were also recorded. Both varieties showed increased grains per panicle by all the applied growth regulators, especially IAA and Kn. In Jaya and Ethrel and Kn in Palman. Similarly, the percentage of the filled grains was also significantly higher with the exogenous application of hormones. All the growth regulators increased the test weight in both varieties over control. Grain yield was maximum with the treatment of IAA in both varieties. The endogenous level of IAA was increased by Kn at both stages in Jaya, whereas in Palman this was only at flowering stage. Gibberellic acid was also increased in Jaya by Kn at the flowering stage. Kinetin and IAA decreased the level of ABA at both stages in both varieties. Gibberellic acid increased IAA only during the panicle emergence stage in Jaya. Reduction in sterility, increased yield and yield contributing parameters were found to be associated with the decrease in the level of growth inhibitor (ABA) by Kn and IAA treatments.

INTRODUCTION

Although environment is the major factor for sterility [1], genetic deficiencies are also thought to be associated with sterility in some varieties of rice [2]. Such problems are common in breeding programmes, specifically radiation breeding [3]. Various reports have brought to light the role of endogenous levels of growth regulators during grain filling of some cereals [4, 5, 2]. The present paper presents the results of studies on the endogenous levels of hormones in two varieties of rice, viz., Jaya and Palman. Very little work has been done to date on the hormonal role with particular reference to the problem of sterility in rice. Substantial yield is lost because of non-filling of the grains, due to such factors as environmental conditions, especially low and high temperatures during the flowering and fertilisation stages, genetic variations and hormonal imbalance at the time of grain development in rice.

MATERIALS AND METHODS

The field experiment was conducted in a randomised block design for both the varieties separately with six treatments of growth regulators. The treatments used were: Control with water spray (T1), IAA 5 ppm spray (T2) GA3 25 ppm spray (T3), Kinetin 10 ppm spray (T4), Ethrel 25 ppm spray (T5) and Control without spray (T6). Spray solutions were prepared according to standard procedures to yield 500 L of solution per hectare. The growth regulators were sprayed at the time of panicle emergence and 50% flowering stages. Samples were collected 5 days after the first spraying and one week after the second spraying. Extraction, purification, bioassays were carried out to determine the levels of hormones from four treatments (T1–T4). Observations were also recorded regarding the effect on yield and yield contributing parameters.

RESULTS AND DISCUSSION

Two varieties of rice, viz., Jaya and Palman were selected; Palman has a known sterility problem that might be due to several factors, of which hormonal imbalance is one. Yield contributing parameters and endogenous levels of hormones were determined to see the effect of exogenous application of some growth regulators in the form of a spray during panicle emergence and flowering stages. Yield and yield contributing parameters were also recorded to see the effect of exogenous application of the growth regulators as shown in Tables 1–4. It can be seen the more detailed investigations are needed to satisfactorily explain the results.

Table 1: Effect of growth regulators on yield and yield contributing characters in Jaya

Treatments	Total No. of grain/p	Percentage of sterile grain	Test wt. (g)	Yield qt/ha
T1: Control (water)	102	25	21.9	75.48
T2: IAA (5 ppm)	131	14	28.0	87.29
T3: GA3 (25 ppm)	120	22	27.2	85.25
T4: Kn (10 ppm)	132	18	28.5	86.65
T5: Ethrel (25 ppm)	115	21	27.1	83.25
T6: Control (Nil)	105	24	21.6	78.22
C.D. at 5%	16.89	–	2.14	7.00

Table 2: Effect of growth regulators on yield and yield contributing characters in Palman

Treatments	Total No. of grain/p	Percentage of sterile grain	Test wt. (g)	Yield qt/ha
T1: Control (water)	99	11	19.1	74.37
T2: IAA (5 ppm)	110	9	21.0	85.84
T3: GA3 (25 ppm)	129	19	20.2	77.47
T4: Kn (10 ppm)	139	22	20.7	77.47
T5: Ethrel (25 ppm)	143	22	20.8	80.06
T6: Control (Nil)	107	16	20.4	72.66
C.D. at 5%-	14.06	–	0.74	4.37

Table 3: Effect of growth regulators on endogenous level of hormones at two stages in Jaya (ng/g fresh weight of sample)

Treatment	Stages	Hormones assayed (ng/g f. wt.)			
		Auxins	Cytokinins	Gibberellins	Abscisic acid
Control (water spray)	Panicle emergence	0.2823	0.8652	0.0061	27.544
	Flowering	0.6392	0.5600	0.0046	53.769
IAA (5 ppm)	Panicle emergence	0.1265	0.4800	0.0042	19.933
	Flowering	0.7600	0.5300	0.0074	1.672
Kinetin (10 ppm)	Panicle emergence	3.2000	0.3500	0.0030	5.406
	Flowering	4.1995	0.4600	0.0267	0.746
GA3 (25 ppm)	Panicle emergence	1.6666	0.5300	0.0061	3.513
	flowering	0.1598	0.3252	0.0252	16.733

Table 4: Effect of growth regulators on endogenous level of hormones at two stages in Palman (ng/g fresh weight)

Treatment	Stages	Hormones assayed (ng/g f. wt.)			
		Auxins	Cytokinins	Gibberellins	Abscisic acid
Control (water spray)	Panicle emergence	1.6666	0.7600	0.0066	5.199
	Flowering	0.3996	0.6600	0.0274	1.413
IAA (5 ppm)	Panicle emergence	0.3599	1.3700	0.0062	8.966
	Flowering	0.3930	0.9960	0.0048	2.689
Kinetin (10 ppm)	Panicle emergence	0.3728	0.3648	0.0018	3.666
	Flowering	0.7264	0.5648	0.0050	0.934
GA3 (25 ppm)	Panicle emergence	0.3796	0.3500	0.0276	19.630
	Flowering	0.3930	0.6900	0.0060	3.918

Nonetheless one may say that all the three growth promoting substances play a role in improving the different yield contributing parameters, though they may function in the sequence that cytokinins play an important role in cell division in grains, followed by gibberellins whose function is cell division and cell enlargement. Auxins may function simultaneously with gibberellins in sequence for accumulation of photosynthates and grain development. In general, endogenous growth-promoting substances were greater during the flowering stage compared to the panicle emergence stage, while the reverse was seen in the case of ABA.

It may be concluded that Palman has no greater sterility problem than Jaya. The investigations for yield contributing parameters and endogenous level of hormones showed very little difference between the two.

LITERATURE CITED

1. Satake, T. 1976. Sterile-type cool injury in rice plants. *Proc. Symp. on Cli. and Rice, IRRI, Philippines*, pp. 281-300.
2. Islam, M.S. 1977. Problem of sterility in rice (*Oryza sativa* L.): Hormonal imbalance between the fertile and sterile lines. *Indian J. Expt. Biol.*, 15: 783-87.
3. Shastry, S.V.S. 1963. Is sterility genic in *Japanica-indica* rice hybrids? In: *Rice Breeding and Genetics* (Proc. Symp. Rice Gen. and Cytogen., IRRI) 154 pp.
4. Dua, I.S. and S.N. Bhardwaj, 1979. Influence of growth regulating substances on grain growth in *Aestivum* wheat. *Indian J. Pl. Physiol.*, 22: 50-55.
5. King, R.W. 1979. Abscisic acid synthesis and metabolism in wheat ears. *Aust. J. Pl. Physiol.*, 6: 99-108.

20

A Study on the Productivity of
Amaranthus hypochondriacus L.

P.K. Jha, J.P. Sah and M.K. Chettri

Central Department of Botany, Tribhuvan University, Kirtipur,
Kathmandu, Nepal

ABSTRACT

The amaranth crop has potential economic value because of its protein quantity and quality. In Nepal, *Amaranthus hypochondriacus* L. is widely cultivated from 60 m to more than 3000 m. Amaranth germplasms, collected from different parts of Nepal and India, were grown under various conditions. Growth parameters were recorded. The yield of grain amaranth was reduced by 69.12% in semi-water-logged conditions. A soybean-amaranth intercrop performed better than monocultures. May to early June is the appropriate time to sow amaranth seeds in the mid-hill climate of Kathmandu valley.

INTRODUCTION

Amaranthus has been identified as a future cereal crop of potential economic value. Amaranth affords a nutritious dish with abundant provitamin A, a vitamin particularly necessary in the tropics for eye health. Its lysine (an essential amino acid) content is also high compared with that found among the most common cereals [1]. Among the three species of grain amaranth (*A. caudatus* L., *A. cruentus* L. and *A. hypochondriacus* L.), the last one is widely cultivated in Nepal while *A. caudatus* is confined to the mountains.

Biologically, amaranth is a C_4 plant which can adapt well in different climates, viz. soil, temperature, light and water. The climatic conditions in

Nepal vary greatly from the south (tropical) to the north (alpine) and here amaranth is grown in the terai (60 m) to an elevation higher than 3000 m. As with other new crops, there are several production uncertainties. Therefore, amaranth productivity was studied under different conditions (semi-water-logged and non-logged, intercropping with legumes and non-legumes and different sowing dates).

MATERIALS AND METHODS

The seeds of different land races of *A. hypochondriacus* were sown in 4 × 4 m plots under the following conditions at Kirtipur Campus, Tribhuvan University during 1986 and 1987.

Waterlogging: The plots were maintained under two water conditions—semi-waterlogged and non-logged. For the first condition, most of the rain water was allowed to remain in the plots; for the latter the rain water was completely drained off.

Intercropping: Amaranth was grown with bean (*Phaseolus vulgaris* L.), soybean (*Glycine max* Mer.) and maize (*Zea mays* L.) in separate plots. The amaranth and partner crop seeds were sown in alternate rows in the plots at a distance of 45 cm.

Varying sowing dates: The amaranth seeds were sown on different dates on interval of one month—last week of April, May and June in 1986 and 1987.

Care was taken under each condition to maintain almost uniform density and the crops were harvested at maturity. Growth parameters, such as plant height, length of terminal and axillary inflorescence, biomass and yield per plant, were recorded at the time of harvest.

RESULTS AND DISCUSSIONS

There are very few records of amaranth productivity under different conditions. Singh [2] found significant variation in amaranth yield from 172 kgha⁻¹ to 1883 kgha⁻¹ in Jumla, Nepal. Joshi [3] reported grain yield of different varieties between 350 kgha⁻¹ to 4100 kg/ha⁻¹ in India.

The growth parameters of different land races of *A. hypochondriacus* varied greatly under different conditions. The crop was adversely affected by waterlogging (Table 1). On average, the yield was reduced by 69.12% and the least affected character was the length of axillary inflorescence. The reduction in other parameters, such as plant height, length of terminal inflorescence and biomass ranged from 41.97 to 57.78%. Of the different land races, the seeds collected from Madhubanni and Lucknow, India gave a poor performance while those from Janakpur, Nepal were comparatively better under water-logging with a yield loss of 32.43%. The experiments at the National Botanical Research Institute, Lucknow showed that amaranth grows better in well-

Table 1: Amaranthus hypochondriacus L. grown in semi-water-logged (A) and non-logged (B) conditions

| Seed Code | Source | Condition | Plant height (cm) | Inflorescence length (cm) | | | Biomass per plant (gm) | Yield per plant (gm) |
| | | | | Terminal | | Axillary | | |
				Main Stalk	Lateral			
A15 Green	Nainital (India)	A	112.20 ±32.34	17.58 ±3.95	5.18 ±2.22	2.58 ±1.63	85.00 ±5.25	29.12 ±5.05
		B	160.00 ±27.77	31.00 ±2.94	11.75 ±1.47	2.52 ±2.25	115.00 ±57.00	88.40 ±20.60
A-15 Red	Nainital (India)	A	95.48 ±39.63	19.57 ±6.30	5.24 ±3.15	4.97 ±3.63	60.00 ±10.50	12.60 ±2.60
		B	147.50 ±14.06	34.00 ±2.94	12.50 ±1.50	2.37 ±1.05	239.00 ±30.50	48.35 ±15.85
A-16	Madhubanni (India)	A	80.92 ±28.53	12.28 ±4.33	2.71 ±1.66	1.85 ±1.72	54.23 ±10.62	20.30 ±5.43
		B	223.33 ±25.60	36.20 ±6.40	12.40 ±2.41	5.05 ±2.37	296.00 ±11.57	76.66 ±2.35
A-20	Lucknow (India)	A	37.14 ±19.16	13.55 ±8.94	5.21 ±3.11	2.01 ±1.70	15.50 ±1.50	0.93 ±0.20
		B	105.00 ±37.20	28.75 ±3.96	12.00 ±4.24	2.50 ±1.50	66.00 ±4.00	13.20 ±2.30
A-21	Lucknow (India)	A	92.57 ±36.25	19.40 ±4.39	8.00 ±3.26	0.83 ±0.37	32.00 ±5.03	5.37 ±1.23
		B	140.00 ±13.78	35.00 ±3.16	9.80 ±2.56	2.50 ±1.25	32.00 ±7.23	21.95 ±8.02
A-23	Dhankutta (Nepal)	A	105.85 ±20.18	13.98 ±4.46	3.24 ±1.37	1.60 ±0.98	102.00 ±25.17	15.85 ±5.63
		B	160.00 ±34.49	38.80 ±6.85	9.90 ±2.69	2.05 ±1.50	180.00 ±15.00	60.05 ±15.17
A-24	Janakpur (Nepal)	A	104.30 ±36.77	30.54 ±12.28	10.45 ±7.43	2.22 ±1.95	141.50 ±8.50	20.65 ±5.23
		B	143.57 ±32.59	22.00 ±7.95	7.33 ±1.37	2.23 ±2.04	233.00 ±26.06	30.54 ±2.89

drained fields. The results confirm that amaranth can grow well, therefore, in relatively dry climates.

Variation in amaranth productivity was also recorded when cropped with bean, soybean and maize (Table 2). The vegetative growth of amaranth was relatively better when intercropped with bean and soybean, the nitrogen-fixing plants. Amaranth grain production was maximum in the soybean-amaranth plot. The marked and significant advantages of corn-legume intercropping over monoculture have been recorded by Rerkasem [4]. The present study strengthens this opinion.

Table 2: Growth parameters of *Amaranthus hypochondriacus* intercropped with bean, soybean and maize

Amaranth with	Plant height	Inflorescence length (cm)		Axillary	Yield per Plant (gm)
		Terminal			
		Main Stalk	Lateral		
Bean	163.28	31.25	8.62	2.57	9.21
(*Phaseolus vulgaris* L.)	± 19.37	± 3.99	± 2.05	± 0.77	± 3.00
Soybean	172.73	37.28	11.62	2.42	13.27
(*Glycine max* Mer.)	± 30.10	± 8.43	± 3.76	± 1.15	± 6.65
Maize	150.87	26.28	8.21	3.85	9.79
(*Zea mays* L.)	± 19.22	± 5.87	± 2.23	± 2.27	± 7.44
Amaranth	142.28	30.14	10.78	3.00	12.82
	± 32.37	± 9.67	± 5.08	± 1.58	± 4.76

Seed sowing dates had a pronounced effect on the productivity of grain amaranth (Table 3). It was best when sown in the last of May. The seeds sown in the last of June had very poor germination and seedling growth due to heavy rain. The seeds sown in the last of April showed poor growth and yielded 61.73% less than the seeds sown in the last of May. The usual practice for amaranth cultivation in the montane regions of Nepal varies from April to June, but in Kathmandu the appropriate time is May to early June after the first rain. In the terai (altitude < 300 m) the cropping season differs, i.e., after the rainy season during October to November. It is concluded that amaranth does better under well-drained conditions, sown with soybean in May to early June in the mid-hill climate of Kathmandu valley.

Acknowledgements

Financial support from the International Foundation for Science, Sweden is thankfully acknowledged. The authors are grateful to Prof. A.R. Shakya and Prof. S.P. Rimal for their encouragement.

Table 3: Effect of seed sowing date on growth parameters of Amaranthus hypochondriacus L.

Seed code	Source	Sowing date	Plant height	Inflorescence length (cm) Terminal Main Stalk	Inflorescence length (cm) Terminal Lateral	Inflorescence length (cm) Axillary	Inflorescence density	Yield per plant (gm)
A-16	Madhubanni (India)	A	101.00 ±16.73	21.39 ±4.80	7.98 ±2.98	2.60 ±1.68	L-I	2.03
		B	158.12 ±32.36	21.14 ±3.35	5.62 ±2.13	5.00 ±3.65	I-D	8.42
A-23	Dhankutta (Nepal)	A	128.32 ±18.44	27.35 ±7.71	8.18 ±3.67	4.77 ±3.01	I-D	3.85
		B	144.07 ±40.49	27.69 ±12.14	6.92 ±3.79	4.18 ±3.22	D	7.75
A-79	Jumla (Nepal)	A	109.50 ±22.89	24.30 ±6.01	6.82 ±4.71	1.64 ±0.85	L-I	3.01
		B	134.42 ±20.80	28.89 ±8.42	6.12 ±3.66	5.30 ±3.61	I-D	5.99
A-84	Jumla (Nepal)	A	129.50 ±20.49	31.08 ±5.10	6.34 ±3.27	2.60 ±1.92	I	2.07
		B	140.50 ±33.30	38.25 ±10.26	8.29 ±4.98	2.67 ±1.33	I	6.46

A = April; B = May last; C = June last (poor germination and seedling growth); L = Lax; I = Intermediate; D = Dense

LITERATURE CITED

1. BOSTID. 1984. *Amaranth: Modern Prospects of an Ancient Crop.* National Academic Press, Washington D.C., USA.
2. Singh, K.M. 1982. Amaranth Field Trials. Paper presented at 10th Summer crop workshop, Nepal. Department of Agriculture, Agronomy Division, Khumaltar, Kathmandu.
3. Joshi, B.D. 1985. Annapurna–A new variety of grain Amaranth. *Indian Farming*, 25 (8): 29-31.
4. Rerkasem, B. 1986. Interactions in Legume-nonlegume intercropping systems. In: *Biological Nitrogen Fixation Program.* Third Coordinating Meeting, BOSTID Research Program) pp. 203-214.

21

SO$_2$ Stress on Yield of *Avena sativa* L.

Subhash Chand and N.K. Yadav

Botany Department, Meerut University, Meerut-250005, India

ABSTRACT

The yield response of *Avena sativa* L. cv. Cannet-2 and cv. Culta was recorded to 0.12 and 0.25 ppm SO$_2$. Ten-day-old seedlings were exposed to 0.12 & 0.25 ppm SO$_2$ for six hours daily in door chambers up to flowering initiation. Foliar injury on margins and tips of both cultivars appeared in 0.25 ppm SO$_2$ after a 20-day exposure and in 0.12 ppm SO$_2$ after a 40-day exposure. Yield in terms of number of grains spike^{-1}, weight of 1000 grains, spike density and harvest index was adversely affected with both concentrations. Flowering was slightly advanced by about 6-8 days in SO$_2$ exposed plants. *A. sativa* L. cv. Cannet-2 was more sensitive than cv. Culta.

INTRODUCTION

Studies of the effect of SO$_2$ pollutant on cereal crops in India are inadequate. The effects of exposure to SO$_2$ can vary despite identical experimental procedures when different plants and cultivars are used [1]. Reduction in growth and yield occurs only after visible injury is apparent. The present study deals with the yield response of *A. sativa* L. cv. Cannet-2 and cv. Culta to SO$_2$.

MATERIALS AND METHODS

Seeds of *A. sativa* L. cv. Cannet-2 and cv. Culta were procured from the National Seeds Corporation, Meerut, and sown in experimental plots, following proper agronomic practices. Ten-day-old seedlings were exposed in 1^{3}m dimension fumigation chambers. The desired concentrations of SO_2 were given with additional flow of air by sodium metabisulphite solution [2]. A control was also run in identical conditions but without SO_2.

RESULTS AND DISCUSSION

To analyse the effects of SO_2, data on foliar injury, sulphur accumulation and yield of plants were recorded. Quantitative observations on morphology and yield were likewise noted. Foliar injury in 60- and 75-day-old leaves was estimated from the amount of the leaf area noticeably altered from its normal morphological form. This injury was usually expressed as a percentage of the leaf area affected by chlorosis or necrosis. Sulphur accumulation in 60- and 75-day-old leaves was determined by the turbidimetric method [3]. Spike density and harvest index were also recorded at crop maturity by dividing the number of spikelets by the length of spike and dry weight of the total grains per plant with dry matter of the whole plant and multiplied by 100.

The first sign of foliar injury appeared after a 20-day exposure in 0.25 ppm SO_2 and in 0.12 ppm SO_2 after a 40-day exposure. Initially, whitish-yellow patches appeared on the margins and tips of the leaves which, on prolonged exposure, turned into dark brown bifacial necrotic lesions.

The leaves of both cultivars showed that leaf area injury increased according to the Sulphur accumulation. The sulphur accumulation and foliar injury (%) in 60- and 75-day-old leaves of *A. sativa* L. cv. Cannet-2 was greater than in cv. Culta (Table 1). Bennett and co-workers [4] found that the amount of leaf destruction related linearly to a product of time vs concentration (dose). Visible injury is associated with cell death, a decrease in total area photosynthetic activity and hence reduced leaf productivity [5].

Table 1: Foliar injury and sulphur accumulation in the leaves of *A. sativa* L. due to SO_2

Parameter	Days	cv. Cannet-2			cv. Culta		
		SO_2 ppm					
		0	0.12	0.25	0	0.12	0.25
Leaf area injury (%)	60	-	13.64	38.71	-	11.40	35.29
	75	-	17.72	43.88	-	15.12	39.45
Sulphur	60	3.75	4.05	4.84	3.68	3.95	4.70
(mg/g dry wt)	75	3.86	4.32	5.36	3.76	4.20	5.16

Plants exposed to SO_2 showed earlier flower initiation than control because plants under stress conditions hurry to complete their life cycle. 0.12 and 0.25 ppm SO_2 caused significant decrease in yield of both cultivars (Table 2). *A. sativa* L. cv. Cannet-2 showed 11.72 and 21.18%, 9.03 and 20.76% reductions in number of grains spike[-1] and weight of 1000 grains with 0.12 and 0.25

ppm SO_2. There was also reduction of 7.79 and 14.53% and 8.79 and 13.85% in number of grains spike[-1] and weight of 1000 grains of *A. sativa* cv. Culta with 0.12 and 0.25 ppm SO_2. Spike density and harvest index of both cultivars likewise decreased significantly.

Table 2: Yield response of *A. sativa* L. to SO_2 pollutant

Parameter	cv. Cannet-2			cv. Culta		
	SO₂ ppm					
	0	0.12	0.25	0	0.12	0.25
Grains spike[-1]	46.82 ± 3.23	41.33* ± 3.11	36.90** ± 2.95	42.46 ± 3.13	39.15 ± 2.98	36.29* ± 2.94
Weight of 1000 grains	30.672 ± 2.91	27.904 ± 2.44	24.308** ± 2.20	29.966 ± 2.85	27.341 ± 2.29	25.810* ± 2.75
Spike density	1.06 ± 0.09	0.97 ± 0.08	0.92* ± 0.08	1.02 ± 0.10	0.95 ± 0.08	0.90 ± 0.07
Harvest index	56.00 ± 4.05	49.18* ± 3.81	45.34** ± 3.72	52.84 ± 3.90	47.54** ± 3.73	43.69** ± 3.57

Values are in mean ± S.D.;

*P < 0.05 and ** P < 0.01 significance difference from control.

Godzik and Krupa [6] reported a 20% loss in economic yield of cereals (wheat, barley, rye and oats) at higher doses of SO_2. Singh and Jain [7] also presented similar effects of SO_2 on yield in *Brassica campestris* cv. Varuna. It is appropriate to comment that percent economic yield loss of both cultivars was less than the percentage leaf area injured and *A. sativa* L. Cannet-2 proved more sensitive than the cv. Culta in terms of yield.

LITERATURE CITED

1. Heagle, A.S. and J.W. Johnston. 1978. Variable responses of soybeans to mixtures of O_3 and SO_2. Air Pollut. Control Assoc., **29**: 241-246.
2. Aggarwal, M., P.K. Nandi and D.N. Rao. 1982. Effects of ozone and sulphur dioxide pollutants separately and in mixture on chlorophyll and carotenoid pigments of *Oryza sativa*. Water Air Soil Pollut., **18**: 49-59.
3. Patterson, G.D. 1958. In: *Colorimetric Determination of Non-metals*. Inter Sc. Publ. Inc., New York., pp. 261-308.
4. Bennett, A.C. Hill and D.M. Gates. 1973. A model for gaseous pollutant absorption by leaves. *J. Air Pollut. Control Assoc.*, **23**: 957-962.
5. Todd, G. 1958. Effect of ozone and ozonated hexane on respiration and photosynthesis of leaves. *Plant Physiol.*, **27**: 435-459.
6. Godzik, S. and S.V. Krupa. 1982. In: *Effects of Gaseous Air Pollution in Agriculture and Horticulture* edited by M.H. Unsworth and D.P. Ormrod. Butterworths, London.
7. Singh, V. and S. Jain. 1987. Effects of SO_2 pollution on *Brassica campestris* cv. Varuna. *Nat. Acad. Sci. Letters*, **10**: (2); 53-55.

22

Investigation on Physiology of Bulb Formation in Some Short-day Onion Varieties

S.K. Regmi*, L. Pun* and I.R. Pandey**

*Vegetable Development Division,
Khumaltar, Kathmandu, Nepal.
**Vegetable Research and Seed Production Centre,
Khumaltar, Kathmandu, Nepal.

ABSTRACT

An experiment was conducted at the Vegetable Seed Production Centre, Mushikot during the 1983 and 1985 seasons to study the physiology of bulb formation in six popular short-day onion varieties in open field conditions. It was apparent from the results that Nasik Red and N-53 produced a greater number of leaves, better plant height (leaf length) and more plant weight by April 21 than the other four varieties, viz., Pusa Red, Red Globe, Red Creole (Nepal) and Red Creole (Northrup King). It was found that the varieties vary in initiation of bulbing and Nasik Red and N-53 were earlier in this respect than the other four varieties.

INTRODUCTION

The Government of Nepal has declared onion as one of the essential commodities in the country. Its cultivation in the kingdom covers more than 1000 ha.

Several scientists have reported the role of leaf area in bulb formation and development [1, 2, 3, 4, 5]. Their studies show that age, size and storage at the start of long-day conditions as well as proper temperature plays very important roles in bulb formation and development. The present study is concerned with short-day varieties, suitable for cultivation in the mid-hills of Nepal.

MATERIALS AND METHODS

The physiology of bulb formation and development was studied in six short-day varieties of onion: Nasik Red, N-53, Pusa Red, Red Globe, Red Creole (Northrup King) and Red Creole (Mushikot). The seeds of these six varieties were sown in a nursery bed on November 24 and the seedlings were transplanted on 24 January (1983 and 1985). The experiment was laid in a Randomised block design. The net plot size was 1.25 m × 5 m with three replications, utilized for bulb yield, while for other observations seedlings were taken from border rows. Manure and fertiliser were applied as recommended for the crop. Farmyard manure, complexal and muriate of potash were applied at the time of soil preparation in individual plots. The crop was top-dressed twice. The bulbs were harvested on 23 June 1983 and 7 June 1985.

Observations on growth and development characters were recorded at 15-day intervals beginning from March 8, viz., development of leaves (number), plant height (cm), plant weight (gm), bulb diameter (cm), and foliage and bladeless scales (number) for each observation, 15 plants were uprooted, washed properly and then examined. To ascertain foliage and bladeless scales the bulbs were dissected and the numbers then counted.

RESULTS

Development of leaves, plant height, plant weight, bulb formation, foliage and bladeless scale are presented in Tables 1–5.

Table 1: Development of leaf (number) average of two years

Sl. No.	Varieties	Dates of observation							
		Mar. 8	Mar. 23	Apr. 7	Apr. 21	May 7	May 22	June 6	June 22
1.	Nasik Red	3.0	4.0	6.0	7.5	8.5	8.5	9.0	9.0
2.	N-53	3.0	4.0	5.5	8.5	9.0	10.5	10.5	9.0
3.	Pusa Red	3.0	3.0	5.5	6.0	8.5	7.5	7.5	8.0
4.	Red Globe	2.0	3.0	5.5	5.5	8.0	7.0	6.0	8.0
5.	Red Creole (North)	2.0	3.0	5.5	6.0	7.5	8.0	7.5	8.0
6.	Red Creole (Mushikot)	3.0	3.0	5.5	6.5	9.5	9.5	8.0	9.0

Table 2: Plant height (cm) average of two years

1.	Nasik Red	9.20	20.75	32.50	41.10	50.45	47.70	49.30	48.00
2.	N-53	16.43	26.10	36.20	39.35	43.60	45.90	57.00	48.20
3.	Pusa Red	4.00	18.40	30.50	53.10	45.25	44.70	41.60	51.80
	Red Globe	10.30	15.80	27.70	34.40	48.95	51.80	48.40	51.00
5.	Red Creole (North)	6.60	12.25	28.05	32.50	44.20	54.80	50.80	52.40
6.	Red Creole (Mushikot)	9.70	22.90	27.90	38.15	47.90	53.20	52.40	52.80

Table 3: Plant weight (gm) average of two years

Sl. No.	Varieties	Dates of observations							
		Mar. 8	Mar. 23	Apr. 7	Apr. 21	May 7	May 22	June 6	June 22
1.	Nasik Red	2.00	4.70	14.00	39.20	79.10	130.30	135.00	135.00
2.	N-53	2.90	3.25	8.70	38.70	60.60	101.30	101.30	117.40
3.	Pusa Red	2.60	1.80	10.15	17.10	53.10	87.60	63.00	92.00
4.	Red Globe	4.00	1.70	7.40	13.90	53.70	84.10	86.00	89.00
5.	Red Creole (North)	3.20	2.15	6.55	12.60	37.70	124.00	78.00	74.00
6.	Red Creole (Mushikot)	2.50	3.45	6.80	17.20	84.60	109.10	86.50	86.00

Table 4: Bulb diameter (cm) average of two years

Sl. No.	Varieties	Dates of observations							
		Mar. 8	Mar. 23	Apr. 7	Apr. 21	May 7	May 22	June 6	June 21
1.	Nasik Red	-	-	1.40	2.60	3.99	5.10	5.10	5.61
2.	N-53	-	-	1.07	2.07	3.51	4.40	5.15	6.41
3.	Pusa Red	-	-	1.16	2.07	2.76	4.27	4.88	5.44
4.	Red Globe	-	-	1.15	1.69	3.22	4.22	4.71	5.56
5.	Red Creole (North)	-	-	1.29	1.37	2.39	4.68	5.04	5.20
6.	Red Creole (Mushikot)	-	-	0.92	1.35	2.63	4.88	4.96	6.51

DISCUSSION

It can seen from Tables 1–5 that all the varieties reached the highest number of leaves by May 7th. However, the Nasik Red and N-53 showed a higher number of leaves by April 21st; the other varieties exhibited rapid leaf development only within the next 15 days. All the varieties had ceased leaf development by May 22. Kato [2] had already proven 25 years ago that severe defoliation of

Table 5: Foliage and bladeless scales (number) average of two years

S. No.	Varieties	Mar 8		Mar 23		Apr 7		Apr 21		May 7		May 22		June 6		June 22	
		FS	BS	FS	BS	FS	BS	FS	BS	FS	BS	FS	BS	FS	BS	FS	BS
1.	Nasik Red	2.8	1.0	3.6	1.2	5.2	1.7	6.8	1.0	10.0	2.0	9.2	3.2	12.0	11.0	9.2	13.0
2.	N-53	2.8	1.0	3.4	1.0	5.5	2.0	6.8	1.8	9.4	2.0	11.0	3.6	15.0	10.0	8.0	16.0
3.	Pusa Red	3.0	1.0	3.2	1.6	4.6	1.2	6.8	1.4	7.8	2.8	9.0	1.8	9.5	10.0	8.8	11.0
4.	Red Globe	2.8	1.0	3.0	1.2	4.8	1.0	6.7	1.4	7.8	1.4	7.5	2.7	7.0	9.0	8.6	15.0
5.	Red Creole (North)	1.8	1.0	3.0	1.6	5.4	1.0	6.4	1.2	8.0	2.4	8.8	1.6	7.4	8.8	7.6	13.4
6.	Red Creole (Mushikot)	2.8	1.0	3.4	2.6	5.0	1.2	6.0	1.8	8.8	1.6	8.6	1.6	7.6	9.2	6.2	12.0

Date of observations

leaves inhibits bulb formation. Terabun [4] subsequently showed that the exposure of just one leaf to continuous light sufficed to induce bulbing on plants with four leaves (the remaining leaves were shielded from light). Heath and Hollies [6] shortly before Kato's publication, had stated that swelling of the seedling section could be induced within a few days even in darkness in the presence of 1 per cent sucrose. This provided a basis for assuming that bulb formation could be triggered only in the presence of a sufficient accumulation of sugars for maintenance of the plant and bulb formation in long-day conditions. Kato [2] tested this assumption and found that carbohydrate accumulation in the leaf sheath reached maximum before the initiation of any bulb development.

Thus it appears that the photosynthetic active leaf area has to produce a sufficient level of sugars before any bulb development can take place a suitable photoperiod. This phenomenon was apparent in the case of the six varieties undertaken in the present study. Of the six varieties, only two, Nasik Red and N-53, produced a greater number of leaves and plant height (i.e., length of leaf) than the other four varieties prior to April 21st. The other four varieties picked up a similar growth pattern within the next 15 days and equalled the plant weight and bulb diameter only by May 22.

However, the Nasik Red and N-53 took a greater number of days for leaf cessation than the other four varieties. It appears that varieties differ in their growth pattern of leaves although the vegetative growth ceased in all the varieties within 4-5 weeks of bulb initiation, as has been reported by Aoba [7]. Interestingly, along with this rapid increase in number, the height also showed a similar growth trend while the plant weight and bulb diameter did not. These later varieties picked up development almost four weeks later than the Nasik Red and N-53. Without doubt the increase in photosynthetic area of the four varieites produced more sugar but increased physiological activities of the plant and longer day conditions delayed the required amount of sugar deposition for the initiation and bulb development process, as has been observed by Kato [2].

From the results of correlation analysis presented in Table 6, it is apparent that a minimum of four leaves is necessary for the initiation of bulb development (plant weight, Bulb diameter). Interestingly, the Nasik Red and N-53 reached the minimum number of leaves as well as greater plant height earlier (by 23rd March), giving more photosynthetically active leaf area for production of the minimum amount of sugar required for bulb initiation. This observation corroborates those of Kato [2], Terabun [4], and Heath and Hollies [6].

At maturity, the onion leaves toppled down due to desiccation of most of the varieties and as a result the plants showed gradual decrease in weight. This stage is generally taken as the final stage of maturity and harvesting of bulbs follows. In fact, at this stage of senescence is translocated into the bulb and activates dormancy. Interestingly, the Nasik Red and N-53 showed no reduction in plant weight. The leaves remained green and even at the harvesting stage leaf initials were noted. It has been reported by Shinde and Sontakke [8] also that these varieties show no senescence, that the vegetative growth continues and hence, expectedly, that the plant weight shows no reduction.

Thus it appears that the varieties Nasik Red and N-53 differ physiologically from the other four varieties, of which Red Creole has been reported to be a 12-hour day type [9]. It is apparent that the Nasik Red and N-53 are shorter day types than the Red Creole as demonstrated hereby the different physiological activities of the varieties under study.

Table 6: Correlation and regression between growth parameters

Independent variables	Dependent variables	Correlation Coefficient	Regression parameters	
			a	b
A. *Nasik Red*				
1. Number of leaves	Plant weight	0.87	23.18	87.43
	Bulb diameter	0.69	1.00	4.2
2. Plant height	Plant weight	0.72	2.60	41.37
	Bulb diameter	0.93	0.75	3.63
B. *Red Creole - Mushikot*				
1. Number of leaves	Plant weight	0.93	18.47	65.07
	Bulb diameter	0.78	0.82	3.15
2. Plant height	Plant weight	0.89	0.16	4.05
	Bulb diameter	0.93	2.54	47.19

Acknowledgements

The authors are grateful to the former and present Chief Vegetable Development Officers, M.N. Pokhrel, and S.B. Aryal respectively, and the Chief Technical Adviser, FAO Fresh Vegetable and Vegetable Seed Production Project, GCP/NEP/043/SWI, S.S. Rekhi for moral support and necessary facilities. The authors especially acknowledge Dr. S.S. Chatterjee, Variety Evaluation and Maintenance Specialist in the above-mentioned FAO Project, Sharad Regmi, Sushila Regmi, P.K. Rai and all other colleges who contributed to the preparation of this paper.

LITERATURE CITED

1. Baker, R.S. and F.E. Wilcox. 1961. *Proc. Amer. Soc. Hort. Science*, **78**: 400-405.
2. Kato, T. 1965. *J. Jap. Soc. Hort. Science*, **34**: 51-57.
3. Vliet, M.V., and J. Scheele. 1966. *Verolagen Van het Landbouwkundig underzoek*, **669**: 27 pp.
4. Terabun, M. 1971. *J. Jap. Soc. Science*, **40**: 50-56.
5. Robinson, J.C. 1971. *Rhodesian J. Agri. Res.*, **9**: 31-38.
6. Heath, .O.V.S. and M.A. Hollies. 1965. *J. Exp. But.*, **16**: 128-44.
7. Aoba, T. 1964. *J. Jap. Soc. Hort. Science*, **33**: 46-52.
8. Shinde, N.N. and M.B. Sontakke. 1986. Bulb Crops. In: *Vegetable Crops in India* edited by T.K. Bose. Naya Prokash, Calcutta, India, pp. 545-581.
9. Shinohara, S. 1977. *Tsukuba Int. Agri. Training Centre, Text Book*, **9**: 38-56.

23

Influence of Night Temperature During Floret Differentiation on Microsporogenesis and Seed-Setting in Sorghum (*Sorghum bicolor* (L.) Moench)

A.M. Dhopte* and J.D. Eastin**

*Reader in Botany, Punjabrao Krishi Vidyapeeth, Akola - 444 104.
**Professor of Agronomy, University of Nebraska, Lincoln, Nebraska, USA-68483.

ABSTRACT

Night temperatures of 17°C, 23°C and 29°C, being near optimum, were imposed during floret differentiation (FD) to study the effect on microsporogenesis and seed-setting in grain sorghum during 1981, under greenhouse conditions. The anatomical studies revealed that stress during FD resulted in PMCs with scanty cytoplasm both in cooler and higher temperature. Cell vacuolation first started in the tepetum. Meiosis was normal in both stresses. In a few cases, the meiotic division was tangential in the stressed plant compared to perpendicular in control. The microspores dissociated from the tepetum under both temp. stresses. At the late tetrad stage, heavy vacuolation was seen in the elevated temperature.

The treatment at FD + 7 hits the post-meiotic stage. Hypertrophy and dilation were, however, the prominent structural changes effected by an elevated temperature. Early degeneration of the tapetum was seen in a cooler temperature. Emptiness in anthers was observed in a cooler temperature. Iodine test indicated 35% and 26% increase in non-viable pollen in cool- and heat-stressed plants respectively. The seed number/plant was significantly reduced in elevated and cooler temp. when stress was imposed at FD (67.7%) and FD + 7 (37.8%) respectively, which paralleled the grain yield reduction by 63.4 and 34.5% respectively.

INTRODUCTION

Temperature stress is regarded as a limiting factor in grain yield when it occurs during the reproductive stage. Floret differentiation has been shown to be the stage most sensitive to night temperature in sorghum [1]. Changes in night temperature beyond optimum (5 to 10°C decreased grain yield in corn and wheat [2] and in sorghum [3, 4]. These yield reductions often paralleled decline in seed number. These results suggest a deleterious effect on panicle development. The adverse effect of temperature or reduction in floret number has been reported [5]. The analysis of cause and effect relationship in grain yield is complex and therefore why yields are reduced is not yet well understood. Floret sterility induced by low temperatures of 15°C or below at the time of meiosis is reported to be a major factor in reducing yields of rice [6]. Low temperatures of 13°C or below during meiosis could induce male sterility in sorghum [7, 8], pearl millet [9] and rice [10, 11]. Microsporogenesis as influenced by night temperature is not well understood. This paper presents the anomalies that occurred during the developmental processes of male gametophytes in response to night temperature.

MATERIALS AND METHODS

A stress-susceptible hybrid of sorghum, RS 671, was planted on 20 August 1982 in 100 plastic pots of 0.011 m capacity filled with soil and peat moss (3:1 v/v). One plant was retained per pot at the 3-4-leaf stage. A few plants were dissected to ascertain the right stage of floret differentiation (FD) (Figs. 1a, 2, 3 and 17). When the plants reached PI + 11 days, they were transferred to three walk-in growth chambers (2.7 × 2.7 × 2.3 m) and were allowed to precondition at temperatures 35/23°C (day/night) for three days prior to the beginning of stress. Night temperature treatments were imposed at FD for 7 days (Figs. 1, 2) and monitored to 17°C and 29°C in two chambers and 23°C for near-optimum in the third chamber on 27 September 1982 (38 days after planting). The PAR at plant height was 800 μE x m^{-2} X s^{-1}. Growth room lighting was provided by four 500-Watt tungsten-halogen lamps + six 100-Watt metal halide lamps RH = 50%). Plants were watered frequently to avoid water stress effects. After seven days, they were removed from the chambers and exposed to normal 35/23°C temperature until maturity. Plants were sampled during the treatment period and the panicles fixed in F.A.A. solution for microscopic examination. Another set of plants was exposed to the same temperature treatment for 7 days at FD + 7 days and the plants again sampled for spikelet examination. The anatomy of the anthers was studied by sectioning the tissue in EPON plastic blocks with an ultramicrotome. Male sterility was studied with 0.1% iodine. Twenty-four treated plants were then kept in the greenhouse under normal conditions. The yield and its components were recorded at harvest and the data were analysed in an FRBD following Steel and Torrie [12].

RESULTS AND DISCUSSION

The studies on microsporogenesis due to cooler and elevated temperature revealed normal meiosis up to the tetrad stage (Fig. 5). Occasionally, division of the sporogenous cell (SC) was not perpendicular to the tapetum. These cells, after tetrad formation, were found dissociated from the tapetum which was vacuolated (Figs. 6, 15). The microspores (MS) were severely shrivelled and dissociated in spite of a normal intact tepetum (Figs. 14, 15). This was followed by shrinkage in the another walls of four locules (Figs. 9, 10). As a result, the pollen grains were pushed to the centre in the locule, resulting in non-dehiscence. The PMCs were full of cytoplasm under normal (23°C) conditions (Fig. 8). The anthers were leaf-like structures (Fig. 9) in the cooler temperature (17°C) but at an elevated temperature (29°C) revealed disorderly development (Fig. 4). There was 25% frequency of empty anthers as against none in stressed plants. Shrinkage in the pollen grains at the late engorgement phase was followed by hypertrophy in the tapetum (Fig. 11) as a result of elevated temperature. Tapetal hypertrophy due to coolness has also been reported for rice by Nishiyama [13] and Sakai [10]. The pollen grains (PG) were rod-shaped and shrivelled with heavy vacuolation in the tapetum. At the time of exine formation, shrinkage in the pollen grains was evident in spite of a normal intact tapetum (Fig. 18) in both the cooler and elevated temperatures (Fig. 16). The cooler temperature not only induced shrinkage in the anther locule with poor development of the connectum (Cn), but the hypertrophied cells of the tapetum pushed the PG into the centre of the anther locule (Fig. 12).

In a few cases the PMCs were devoid of cytoplasm and did not differentiate in the sporogenous cell (Fig. 13). These responses were mostly common in both the cooler and elevated temperature imposed during FD and FD + 7 with minor variation. The data on male sterility, as observed by the iodine test, revealed a significant increase in male sterility by 46% in cooler and 60% in elevated temperature imposed at FD only (Table 1). Elevated temperature did not however elicit a significant response at FD + 7. These data accord with that of Dowens and Marshall [7] and Brooking [8] due to low temperature and Gonzalez [14] due to high temperature in sorghum. Male sterility due to low temperature is also reported in pearl millet [15] and in rice [11, 6]. The higher temperature increased male sterility in corn [16], wheat [17] and rice [18].

An elevated temperature of 29°C at FD significantly reduced the seed number by 67.7% (Table 2) while a reduction of 37.8% was seen at FD + 7 in the cooler temperature (17°C). Sorghum is protandrous and hence pollen development is affected first. It appears sensitive to a cooler temperature later, at FD + 7, when megasporogenesis begins. These responses corroborate the findings of various workers in cereals [1, 19] regarding an elevated temperature. Contrarily, Tingle and co-authors [20] reported an increase in floret number

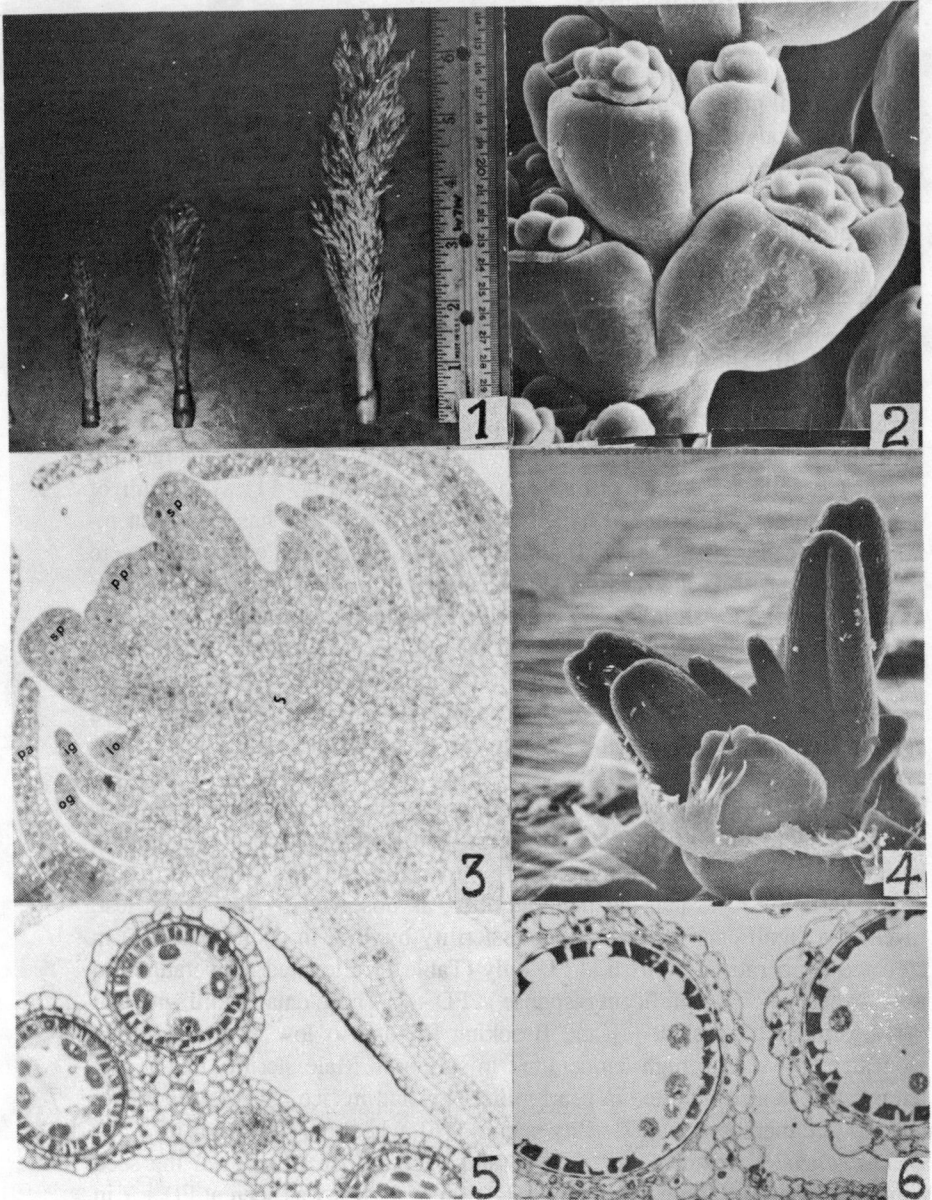

Figs. 1–6: 1a—Floret differentiation (FD); b—fd + 2, c—FD + 4, d—FD + 7; 2—Enlarged
view of floret differentiation, a SEM Micrograph showing three anthers and one
pistil in the centre; 3—Floret anatomy of Fig. 2; 4—Disorderly development of
anthers; 5—Anatomy of anther showing two locules with tapetum having full cytoplasm; 6—Dissociation of microspore (MS) from intact tapetum.

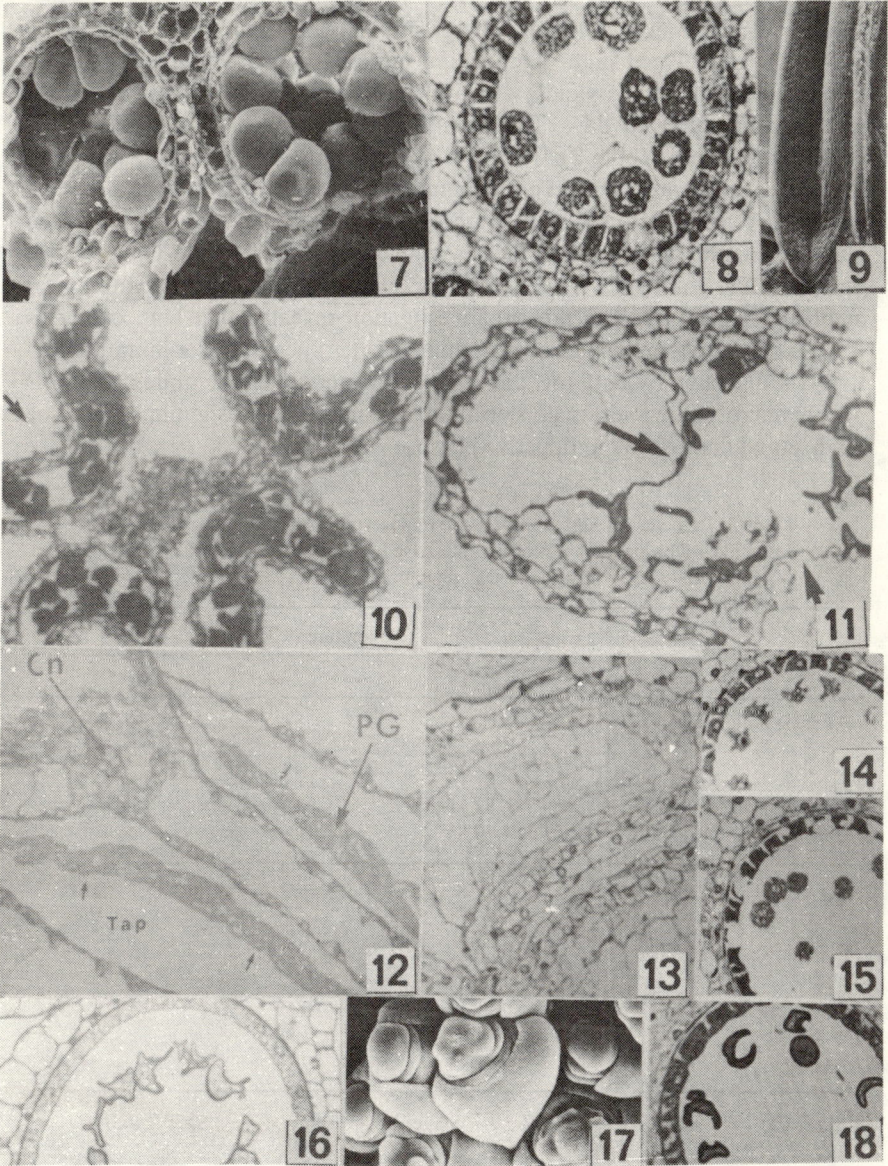

Figs. 7–18: 7—SEM of an anther showing two locules with pollen grains 610 X; 8—Micro-spore at late prophase under normal temperature 400 X; 9— Anthers with leaf-like structure in cooler temperature 68 X; 10—Shrinkage in anther locule 375 X; 11— Hypertrophy in tapetum (arrow) 376 X; 12—Rod-like pollen grains (PG) in shrivelled locule 394 X; 13—Undifferentiated tissue in anther locule in cooler temperature (17°C) 373 X; 14 Shrinkage in microspore 370 X; 15—Dissociation of MS from tapetum 323 X; 16—Shrinkage in PG at engorgement stage with normal locule wall in elevated temperature (29°C), 376 X; 17—SEM of floret prior FD 135 X; 18—Abortion of PG with intact normaltapetum 376 X.

due to low temperature in barley. The grain yield was significantly reduced by 63.4% by an elevated night temperature imposed at FD. However, a cooler temperature reduced the yield by 34.5% from non-stressed plants when stress was imposed at FD + 7 days. The grain yield reduction was paralleled by seed number reduction. (Table 2). Thus, FD appears most sensitive to temperature stress. A good parallel between seed number and grain yield indicates that the treatment effects on grain yield were primarily due to their effects on seed number. Gomez [21] reported 13.7% reduction in yield while Ogunlela [1] observed 28% grain yield reduction by 5°C elevation in night temperature over ambient temperature. The present investigation revealed that both cooler as well as elevated temperature over optimal are detrimental to sorghum productivity. A deviation by only 6°C above or below normal temperature during FD affects microsporogenies, male sterility and grain yield in sorghum by reduced seed-setting during the reproductive phase.

Table 1: Influence of night temperature imposed at Floret Differentiation (FD) and FD + 7 on per cent pollen fertility based on iodine test in RS 671 grain sorghum. Day temperature, 35°C

| Stage | Temperature, °C | | | |
	17	23	29	Mean
(a) Fertile pollen (%):				
FD	17.35*	67.14	17.04*	33.84
FD + 7	63.54*	82.56	80.79	75.63
Mean	40.45	74.85	48.91	-

LSD (0.05) = 16.57,   CV = 20.4%,   Stage**,   Temp.**,   SXT.**   (0.01)

Stage	17	23	29	Mean
(b) Semifertile pollen (%):				
FD	28.81*	24.83	15.17*	22.92
FD + 7	3.03	3.01	3.08	3.04
Mean	15.92	13.92	9.11	-

LSD (0.05) =   3.39,   CV = 37.0%,   Stage **,   Temp. **,   SXT ** (0.01)

Stage	17	23	29	Mean
(c) Sterile pollen (%):				
FD	53.83*	8.28	67.81*	43.31
FD + 7	24.43*	12.91	16.07	17.81
Mean	39.13	10.60	41.94	-

LSD (0.05) = 4.68,   CV = 21.7%,   Stage**   (0.01P),   Temp**,   SXT**

Table 2: Influence of night temperature imposed at Floret Differentiation (FD) and FD + 7 on seed number/panicle and grain yield/plant (g) in RS 671 grain sorghum.

| State | Day/Night Temperature Treatment °C | | | |
	35/17	35/23	35/29	Mean
	Seed number/panicle			
FD	428.75	726.0	234.75*	463.17
	(-40.9)	-	(-67.7)	
FD + 7	420.0	675.0	514.75	536.58
	(-37.8)	-	(-23.7)	
Mean	424.38	700.50	374.75	

LSD (0.05) = 226.85, CV = 30.55%, Stage = NS, Temperature = **, S×T = NS.

	Grain yield/Plant (g)			
FD	13.75	16.58	6.01*	12.11
	(-17.1)	-	(-63.4)	
FD + 7	12.21*	18.64	15.58	15.48
	(34.5)	-	(-16.4)	
Mean	12.97	17.61	10.79	-
	(-26.4)	-	(-38.7)	

LSD (0.05) = 5.70, CV = 27.8%, Stage *, Temp. ** (0.01); S×T *

Figures in parentheses indicate percent reduction from control.

LITERATURE CITED

1. Ogunlela, V.B. 1979. Physiological and agronomical responses of a grain sorghum (*Sorghum bicolor* (L.) Monech) hybrid to elevated night temperature. Ph.D.Diss., University of Nebraska, Lincoln, Nebraska, USA.
2. Peters, D.B., J.W. Pendleton, R.H. Hageman and C.W. Brown. 1971. Effects of night or air temperature on grain yield of corn, wheat and soybean. *Agro. J.*, **63**: 809.
3. Dowens, R.W. 1972. Effect of temperature on phenology and grain yield of *Sorghum bicolor* (L.) Monech. *Aust. J. Agric. Res.*, **23**: 585-594.
4. Eastin, J.D., R.M. Castlebery, T.J. Gerik, J.H. Hultquist, V. Mahalaxmi, V.B. Ogunlela and J.R. Rice. 1983. Physiological aspects of high temperature and water stress. In: *Crop Relations to Water and Temperature Stress in Humid Temperature Climate*. edited by David Raper, Jr. and P.J. Kramer. Westview Press, Boulder, Colorado, USA.
5. Mohapatra, P.K., D. Aspinal and C.F. Jenner. 1983. Differential effects on floral development and sucrose content of the shoot apex of wheat. *Aust. J. Plant Physiol.*, **10**: 1-7.
6. Board, J.E. 1979. Low temperature floret sterility in rice as affected by cultivar and environmental characteristics. *Diss. Abst. Inter.*, B 39 (12): 56-95.
7. Dowens, R.W. and D. R. Marshall. 1971. Low temperature induced male sterility in *Sorghum bicolor* (L.) Moench. *Sorghum Newsletter*, **14**: 11-12.
8. Brooking, I.R. 1976. Male sterility in *Sorghum bicolor* (L.) Moench induced by low night temperature. I. Timing of Stage of sensitivity. *Aust. J. Plant Physiol.*, **3**: 589-596.

9. Mashingaidze, K. and S.C. Muchena. 1982. The induction of floret fertility by low night temperature in pearl millet (*Penisetum typhoids* (Burn) Stapf and Hubbard. *Zimbabwe J. Agric. Res.*, **20**: 29-37.

10. Sakai, K. 1949. Cytohistological, thermotrological studies of sterility of rice in northern parts of Japan with special reference to abnormal hypertrophy of tapetal cells due to low temperature. Rep. *Hokkaido Agri. Expt. Sta.*, **43**: 1-46.

11. Satake, J. 1976. Determination of most sensitive stage to sterile type cool injury in rice plants. *Res. Bull, of Hokkaido Nat. Agri. Expt. Sta.*, **113**: 1-44.

12. Steel, R.GD and J.H.Torrie. 1980. *Principles and Procedures of Statistics. A. Biometrical Approach.* McGraw-Hill Inc., N.Y. (2nd ed.)173 pp.

13. Nishiyama, I. 1970. Male sterility caused by cooling treatment of young microspore stage in rice plant. VIII. Electron microscopical observation on tapetal cells dilated by cooling treatment. *Proc. Crop. Sci. Soc. Japan*, **39**: 480-486.

14. Gonzalez-H, V.A. 1982. Sorghum responses to high temperature and water stress imposed during panicle initiation. Ph.D. Diss. University of Nebraska, Lincoln, Nebraska, USA, 108 pp.

15. Rao, M.K. and Umadevi. 1983. Variation in expression of genic male sterility in pearl millet. *J. Heredity*, **74**: 34-38.

16. Herrero, M.P. and R.R. Johnson. 1980. High temperature and pollen viability of maize crop. *Crop Sci.*, **20**: 796-800.

17. Saini, H.S. and D. Aspinal. 1982. Abnormal sporogenesis in wheat induced by short periods of high temperature. *Ann. Bot.*, **49**: 835-845.

18. Satake, T. and S. Yoshida. 1978. High temperature induced sterility in Indica rices at flowering. *Japan J. Crop Sci.*, **47**: 6-17.

19. Eastin, J.D. 1976. Temperature influence on sorghum yield. *Proc. 31 Ann. Corn and Sorghum Res. Conf. Amer. Seed Trade Assoc., Washington D.C.* edited by H.D. Loden and D. Wilkinsen, pp. 19-23.

20. Tingle, J.N., D.C. Faris and D.P. Ormrod. 1970. Effects of temperature and light and variety in controlled environments on floret number fertility in barley. *Crop. Sci.*, **10**: 26-28.

21. Gomez, J. 1982. Effects of water stress and night temperature on grain yield of *Sorghum bicolor* (L.) Moench under field conditions. MS thesis, University of Nebraska, Lincoln, Nebraska, USA.

24

Changes in Total Soluble Protein and Leghaemoglobin Content of the Nodules at Various Stages of Plant Growth in Relation to Flowering and Pod Formation in *Vigna sinensis* L.

Nisha Thakral

Botany Department, Hindu College, Moradabad - 244 001 (India)

ABSTRACT

Studies were done to ascertain the changes in the total soluble protein and leghaemoglobin content of the nodules with increasing age of field-grown plants under normal environmental conditions. The nodule number increased up to the age of 50 days after germination (DAG) and thereafter decreased up to 65 days. The fresh weight of the nodules increased with the age of the plant and reached maximum at 45 DAG. Initiation of nodule senescence was likewise recorded at 45 DAG. The maximum total soluble protein was noted at 40 DAG and leghaemoglobin content was maximum at 45 DAG. The first flowering was observed on the 43rd and pod formation was recorded on the 46th DAG.

INTRODUCTION

The symbiotic association of a legume with rhizobia, resulting in the formation of nodules, constitutes one of the few natural process leading to the conversion of unusuable atmospheric nitrogen into a form that can readily be utilised by plants. The formation of nodules in grain legumes is related to the flowering time. It has been reported that in chickpea delayed flowering conditions increase nodulation [1].

This paper presents the results of studies on the changes that occur in nodulation and nodular protein with increasing age of plant and also in relation to flowering, pod formation and seed formation.

MATERIALS AND METHODS

Plants were raised in the field under normal environmental conditions. Throughout the experiment the crop experienced an average minimum temperature of 23⁰C and an average max. temperature of 38.3⁰C. Common practices were followed in the preparation of the field and seed sowing. Irrigation was done from time to time, as required. Plants were uprooted on different days after germination, i.e., at 10, 15, 20, 25, 30, 35, 40, 45, 50, 55, 60, 65 days, with intact nodules on the roots. The dates of the first flowering and pod formation were recorded. The nodulation parameters recorded were: nodule number, nodule mass and initiation of nodule senescence. The nodules were carefully removed from the roots on an ice mixture and used in estimating the total soluble protein and leghaemoglobin (Lb) content. The nodules were washed in prechilled and sterilised distilled water and then crushed in a Tris Buffer (pH 9.9). The slurry was centrifuged at 1000 × g for 10 minutes. The supernatant was made up to a known volume and was used for the estimations. The Lb content was estimated by using the method of Hartee [15] with a slight modification—O.D. was taken at both 535 nm and 556 nm. The total soluble protein was estimated as per the Bio-Rad's Technical Bulletin [16] against BSA calibration curve and the O.D. was taken at 595 nm.

RESULTS

The number of nodules increased with increasing age of the plant from 10 to 50 days after germination (DAG); thereafter the number reduced significantly. The initiation of nodule senescence was recorded on the 45th day of germination. The first flowering was observed on the 43rd DAG and pod formation on the 46th day. Seed formation occured on the 50th day of germination. The fresh weight of the nodules followed the same pattern observed for number of nodules. The fresh weight of the healthy nodules increased up to the 45th day. Reduction in fresh weight of nodules was recorded from the 50th to the 65th DAG due to senescence of the nodules; on the other hand the fresh weight of senescent nodules increased from the 45th to the 65th DAG.

The total soluble protein increased from the 10th to the 40th DAG and thereafter reduced significantly, while the Lb content increased with increasing age of the plant and reached maximum on the 45th day of germination. In between the 35th and 40th day and the 40th and 45th day of germination, the

Lb content changed very little. However, from the 50th to the 65th day a remarkable reduction was noted in the Lb content of the nodules.

DISCUSSION

It is clear from the foregoing observations that after the first flowering the nodulation parameters, total soluble protein and Lb content start to decrease. This shows a clear relationship between the flowering and nodulation. The reduction in nodulation parameters is due to the lesser availability of photosynthates to the nodules [2]. Streeter [3] has suggested that some factors or metabolites other than carbohydrates, shared by nodules and other plant parts, are responsible for the decline in nodule activity. A common pattern of nitrogen fixation has been observed in the few legumes studied to date. The nitrogen fixation initially increases during vegetative growth, becoming maximum at flowering, then declines as the plant matures [4, 5, 6, 7, 8]. The same pattern was observed in the present study for total soluble protein and leghaemoglobin content of the nodule. Studies done on root respiration per unit of fixed nitrogen in the cowpea showed that it was maximum at the time of flowering [9]. The change in nodule respiration also tended to parallel changes in root respiration and the amount of nitrogen fixed [10]. The assimilates produced in the leaves after flowering were mainly transported to growing pods, which are the major sinks rather than the nodules [11, 12, 13, 14]. These results indicate that an inadequate supply of photosynthates to the nodules initiates nodule senescence as well as a reduction in nodule number, nodule mass and concomitantly nodular proteins.

LITERATURE CITED

1. Dart, P.J., R. Islam and A. Eaglesham. 1975. The root nodule symbiosis of chickpea and pigeon pea. *International Workshop on Grain Legumes* (ICRISAT), Jan. 13-16, pp. 63-83.
2. Sheoran, I.S., Y.P. Luthra, M.S. Kuhad and Randhir Singh. 1981. Effect of water stress on some enzymes of nitrogen metabolism in pigeonpea. *Phytochemistry*, **20**: 2675-2677.
3. Streeter, J.G. 1981. Seasonal distribution of carbohydrates in nodules and stem exudates from field-grown soybean plants. *Annals of Botany*, **48**: 441-450.
4. Lawn, R.J. and W.A. Brun. 1974. Symbiotic nitrogen fixation in soybean. I. Effect of photosynthetic source-sink manipulations. *Crop Sci.*, **14**: 11-16.
5. Herridge, D.G. and J.S. Pate. 1977. Utilization of net photosynthate for nitrogen fixation and protein production in an annual legume. *Plant physiol.*, **60**: 759-764.
6. Bethlenfalvay, G.J., S.S. Abu-Shakra, K. Fishbecj and D.A. Phillips. 1978. The effect of source-sink manipulation on nitrogen fixation in peas. *Plant physiol.*, **43**: 31-34.
7. Van Berkum, P. and C. Sloger. 1981. Ontogenetic variation in nitrogenase, nitrate reductase and glutamine synthase activities in *Oryza sativa*. *Plant physiol.*, **68**: 722-726.
8. Young, J.P.W. 1982. The time course of nitrogen fixation, apical growth and fruit development in peas. *Annals of Botany*, **49**: 135-139.
9. Neves, M.C.P., F.R. Minchin and R.J. Summerfield. 1981. Carbon metabolism, nitrogen

assimilation and seed yield of cowpea plants dependent on nitrate nitrogen or one of the two strains of *Rhizobium Tropical Agriculture*, **58**: 115-132.

10. Luthra, Y.P., I.S. Sheoran and Randhir Singh. 1983. Ontogenetic interactions between photosynthesis and symbiotic nitrogen fixation in pigeonpea. *Ann. Appl. Biol.*, **103**: 549-556.

11. Latimore, M., J. Giddens and D.A. Ashley. 1977. Effect of ammonium and nitrate nitrogen upon photosynthate supply and nitrogen fixation by soybean. *Crop Sci.*, **17**: 399-404.

12. Hume, D.J. and J.G. Criswell. 1973. Distribution and utilization of ^{14}C-labelled assimilation in soybean. *Crop Sci.*, **13**: 519-524.

13. Kuo, C.G., M.C.H. Jung and S.C.S. Tsou. 1978. Translocation of ^{14}C photosynthate in mungbean during the reproductive period. *Horticulture Science*, **13**: 580-581.

14. Pearen, J.R. and D.J. Hume. 1981. ^{14}C-labelled assimilate utilization by soybean grown with three nitrogen sources. *Crop Sci.*, **21**: 938-942.

15. Hartee, 1955.

16. Bio-Red. Tech. Bull., (1977).

25

Cytological Studies on
Nitrous-oxide Treated Red Clover
(*Trifolium pratense* L.)

N. Giri

Central Department of Botany, Tribhuvan University,
Kathmandu, Nepal

ABSTRACT

An autotetraploid red clover (*Trifolium pratense* L.) population derived by nitrous oxide was examined to see the effect of nitrous oxide and the frequency of euploids and aneuploids produced. A comparative study of diploids and tetraploids was also done. Seven bivalents were observed in the PMCS of diploid red clover. The somatic chromosomes in the diploids and tetraploids (euploids) were 2n = 14 and 2n = 28 respectively. Some aneuploids were also observed. Meiotic studies of tetraploids showed the presence of univalents and multivalents. It was confirmed that nitrous oxide is a good agent for producing more euploids than aneuploids in red clover.

INTRODUCTION

Trifolium pratense L. commonly known as red clover whose tetraploid varieties are reported to be more disease-resistant and longer-lived than diploids. Tetraploids can be produced in red clover by colchicine treatment and the unreduced gamete method [1] the latter involves doubling of reproductive tissue.

Nitrous oxide treatment has been applied by Berthaut [2, 3], Matsu-ura and co-workers [4], Taylor and co-workers [5] in red clover. Ellerstrom and Sjodin [6] found 35% aneuploidy in colchicine-derived tetraploid red clover.

The objectives of the present work were: (1) a comprative study of diploid and tetraploid red clover both morphologically and cytologically; and (2) determinations of the frequency of anenploids and euploids in nitrous-oxide derived tetraploids.

MATERIALS AND METHODS

Seeds of both diploid and tetraploid (N_2O derived) red clover were received from the department of agronomy, University of Kentucky, Lexington, U.S.A. Seeds were grown to the flowering stage in the pots at the Botany Department, (T.U.), Kirtipur, Kathmandu, Nepal.

METHODS

Mitotic studies were made on the root tips. The root tips were pretreated at 0.003 8-hydroxyquinoline for 6 hours, fixed in alcohol acetic acid (3:1/v.v) and stained in Feulgen. A chromosome number count was done and the frequency of tetraploids and aneuploids produced also observed.

Flower buds for chromosome count and meiotic behaviour were fixed in alcohol-acetic acid (3:1/v.v) and stained in acetocarmine. Pollen viability was checked by staining in acetocarmine. Diploids and tetraploids were compared on the basis of morphological and cytological studies.

RESULTS AND DISCUSSION

When the tetraploids were examined for chromosome number count, some euploids and aneuploids (2n = 27, 29 and 30) were observed. The frequency of aneuploids produced was 46%. This aneuploid percentage is similar to that obtained by Maizonnier and Picard [7], Maizonnier [8] and Ellerstrom and Sjodin [6], who reported an aneuploid frequency ranging from 35 to 50%. The advantage of nitrous oxide treatment lies in the higher frequency of tetraploids and the absence of mixoploids. Mitosis in the diploid was also observed. The somatic chromosome number was 2n = 14. So the production of more euploids than aneuploids is the result of nitrous oxide treatment. Meiosis in both diploids was normal (n = 7) while the tetraploids showed the presence of univalents and multivalents. At anaphase I generally 14 chromosomes were observed at each pole but sometimes chromatin bridges were also seen (Fig. 1 and Table 1).

Pollen grain stainability was 90% in diploids but 80% in euploids (2n = 28) and 56-57% in aneuploids. The higher rate of pollen sterility in tetraploids might be due to an inversion bridge, causing deficiency and genic imbalance in the microspores.

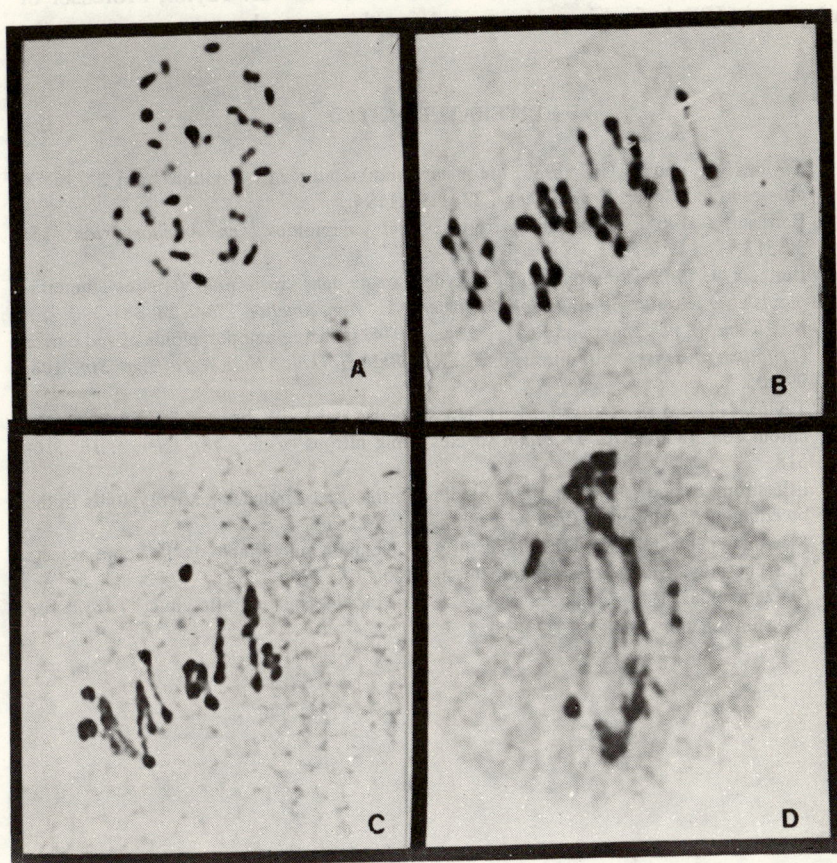

Fig. 1: Mitotic and Meiotic cells of autotetraploid red clover (*T. pratense*) A—Mitotic metaphase with 2 n = 29 X 2100; B—meiotic metaphase with n = 14 X 2100; C—meiotic chromosomes with univalents and multivalents X 2100; D—chromatin bridge at anaphase I X 2100.

Table 1: Irregularities in disjunction at anaphase I of euploids and aneuploids from a nitrous-oxide-derived tetraploid population of red clover

Chromose No.	No. of plants	No. of PMCs	Lagging chromosomes		Cells with bridges
			Percent cells exhibiting	Range per cell	
28	2	33	30.30	0 to 3	1
29	1	70	11.51	1 to 2	0
30	1	20	30.00	0 to 2	0

Acknowledgements

The author expresses her sincere thanks to Dr. N.L. Taylor, Professor of Agronomy, U.S.A. for his guidance in doing this research work.

LITERATURE CITED

1. Taylor, N.L. and N. Giri. 1983. Frequency and stability of tetraploids from 2X × 4X crosses in red clover. *Crop Science*, **23**: 1191-1194.
2. Berthaut, J. 1965. Obtention de trefles violet tetraploides. *Ann. Amelior. Plant*, **15**: 37-51.
3. Berthaut, J. 1968. L' emploi du protoxyde d'ozote dans creation de varieties autotetraploids chez le trefle violet (*Trifolium pratense* L.) *Ann. Amelior. Plant*, **28**: 381-390.
4. Matsu-ura, M., Y. Maki and R. Hayakawa. 1974. Inducing autopolyploids of red clover (*Trifolium pratense*) with nitrous oxide. *Res. Bull. Hokkaido Natl. Agric. Exp. Stn.*, **108**: 99-105.
5. Taylor, N.L., M.K. Anderson, K.H. Quesenberry and L. Watson. 1976. Doubling the chromosome number of *Trifolium* species using nitrous oxide. *Crop science*, **16**: 516-518.
6. Ellerstrom, S. and J. Sjodin. 1974. Studies on the use of induced autopolyploids in the breeding of red clover. *Z. Pfanzenzuchtung*, **71**: 253-263.
7. Maizonnier, D. and Picard. 1970. Fertility problems in autotetraploids. II. Some aspects of aneuploidy. *Ann. Amelior. Plant*, **20**: 407-420.
8. Maizonnier, D. 1969. Quelques aspects de la triploidie chez le trefle violet (T. *pratense* L.) Ann. *Amelior. Plant*, **19**: 277-288.

26

Pollen Studies in *Cardamine* L. (Cruciferae Juss.)

K.S. Khatri

Department of Biology, S.N. Campus, Mahendra Nagar,
Kanchanpur, Nepal

ABSTRACT

Based on herbarium material, the pollen morphology of 32 species of Cordamine (Cruciferae) was studied from the Asian territory of the Soviet Union. This study was carried out with LM and SEM. Taxonomic variations were found in pollen diameter, exine thickness, texture (size of exine lumina), number and breadth of colpi, which are characteristic of species or groups of species; however, these characters are not correlated taxonomically at the section level.

INTRODUCTION

Polynological studies in Cruciferae have been carried out by many authors [1, 2, 3, 4, 5, 6, 7, 8, 9, 10, 11, 12, 13, 14, 15]. Pollen grains have generally been regarded of limited taxonomic value in Cruciferae because of their uniform morphology. But some authors have remarked the importance of pollen characters in predetermining the relationships among genera and suprageneric taxa [4, 7, 8, 9]. Avetisian [16] on the other hand, thinks that these characters can be used only in tracing out the relationships of controversial intra- and interspecific and generic taxa.

After a detailed survey of the pollen morphology of Cruciferae, Chiguri-aeva [7, 8, 9] demonstrated variations in pollen diameter, exine thickness, reticulation pattern, breadth and number of colpi which are characteristic of species or genera or suprageneric taxa. According to her, primitive taxa are characterised by small, finely reticulate pollen and the advanced ones by larger, coarsely reticulate grains.

MATERIALS AND METHODS

The present study covers the pollen morphology of 32 species of *Cardamine* from the Asian territory (including the Coucasus) of the Soviet Union. Material was collected from dried herbarium specimens at LE, LECB, MHA, MW & MOSM. Pollen grains were studied with the LM in fuchsin-glycerine and after acetolysis [1, 24]. For each species 2-4 samples were analysed from different parts of their ranges and in each case the average pollen diameter is based on ca. 100 counts. For greater details of surface features non-acetolysed, dry pollen grains were studied with the SEM (JSM-35c model).

The results of pollen measurements are given in Table 1. The taxonomic arrangement is based on Schulz [17] with some modifications and additions [13, 15]. In each section the species are arranged alphabetically, irrespective of their phylogenetic relationships. The chromosome numbers are based on the literature [see 18, 19, 20, 21].

RESULTS AND DISCUSSION

The dry pollen grains are ellipsoidal. However, when treated with fuchsin-glycerine or acetolysed, they appear spheroidal or sometimes oblate-spheroidal. But in some populations the latter form predominates. This type of behaviour may be attributed to allelic segregation [22, p. 30]. Size 15.5-37.2 µm, 3(4) colpate, colpi 3.2-8.5 µm wide, fairly long running almost throughout the polar axis, with tapering ends and smooth membrane. Exine 0.5-3.5 µm thick, thicker at the centre of the mesocolpium, decreasing in thickness towards the colpi and polar regions. Sexine thicker than nexine, consisting of a columellar layer of cylindrical rods which unite on the surface to form a finely or coarsely reticulate texture that breaks in the region of the colpi. The nexine is thin and smooth (Figs. 1-3).

As demonstrated previously [8, 11, 13, 14, 15 22, 23,], the pollen grains show some variations in diameter, exine thickness, texture (size of exine lumina), breadth and number of colpi, as shown in Table 1. But pollen characters do not correlate with floral ones and do not seem useful in delimiting the sections. But on the basis of size [1] and surface features [8] they can be grouped as follows:

1. Pollen small sized (diameter less than 25.0 µm) and finely reticulate (exine lumina 0.4-2.5 µm).
2. Pollen medium sized (diameter exceeding 25.0 µm) and finely reticulate.
3. Pollen medium sized and coarsely reticulate (exine lumina 1.5-4.0 µm).

Table 1: Measurements of pollen grains of *Cardamine* (in μm)

Sections	Species	Diameter		Exine thickness	Size of Breadth		No. of colpus	2n
		Average	min.-max.		exine	lumina		
1. Sect. *Sphaerotorrhiza* Schulz								
	1. *C. trifida* (Poir.) Jones	24.7-27.2	22.3-29.2	2.0-2.8	1.5-4.0	4.5-6.5	3(4)	32,48
2. Sect. *Macrophyllum* Schulz								
	2. *C. amaraeformis* Nakai	28.1-29.1	26.3-31.2	2.4-3.0	1.6-2.2	6.0-7.5	3(4)	-
	3. *C. densiflora* Gontsch.	26.6-28.0	20.0-30.0	1.6-2.5	1.0-1.8	5.0-6.5	3(4)	-
	3. *C. leucantha* (Tausch) Schulz	18.3-19.8	16.6-21.2	1.2-1.8	1.2-2.0	4.0-5.0	3	16,24
	4. *C. macrophylla* Willd.	24.2-28.6	22.3-31.2	1.6-2.8	1.8-2.5	5.0-7.0	3(4)	50, 64, 80, 96
	5. *C. prorepens* Fisch. ex DC.	24.1-29.5	22.3-31.2	2.0-2.8	1.4-2.2	5.5-7.5	3(4)	-
	6. *C. schinziana* Schulz	19.9-21.7	17.8-23.8	1.4-1.8	1.2-1.8	4.2-5.0	3	-
	7. *C. tomentella* (Vorosh.) Shlothg.	21.9-23.3	20.5-24.5	1.4-2.0	1.2-2.0	4.3-5.0	3(4)	16, 32, 46-48, 72
	8. *C. yezoensis* Maxim.	23.3-28.0	20.5-29.7	1.4-2.4	1.5-2.2	4.5-6.0	3(4)	-
3. Sect. *Lyratum* Khatri								
	9. *C. fauriei* Franch.	23.6-24.5	20.0-26.2	1.0-1.6	0.4-0.8	4.0-5.0	3	-
	10. *C. lyrata* Bunge	20.3-21.3	19.2-22.7	1.0-1.6	0.4-0.8	4.0-5.0	3	-
4. Sect. *Cardamine*								
	11. *C. acris* Griseb.	24.4-25.5	21.5-27.4	2.0-2.7	1.8-2.8	4.5-6.0	3	12
	12. *C. amara* L.	17.6-18.8	15.6-21.2	1.3-1.8	1.0-1.6	4.0-5.0	3	16, 24, 32
	13. *C. dentata* Schult.	26.1-30.9	24.5-34.3	2.8-3.5	2.5-4.0	4.8-6.2	3(4)	56-96, 118
	14. *C. hirsuta* L.	23.5-25.7	22.3-27.8	1.8-2.4	2.5-4.0	4.5-6.2	3	16
	15. *C. impatiens* L.	17.6-18.8	16.4-21.2	0.6-1.0	0.6-1.2	3.5-4.5	3	16

Contd.

Table 1. Contd.

Sections	Species	Diameter min.-max.	Diameter Average	Exine thickness	Size of Breadth exine	Size of Breadth lumina	No. of colpies	2n
	17. *C. nymanii* Gand.	26.7-37.2	27.5-33.3	2.2-2.9	2.0-3.0	4.5-6.0	3(4)	56-100
	18. *C. parviflora* L.	17.8-21.2	18.5-19.5	1.7-2.4	1.5-2.4	3.5-4.5	3	16
	19. *C. pectinata* Pall. ex DC.	17.4-22.3	19.0-20.5	0.5-1.0	0.4-1.0	4.0-4.8	3	-
	20. *C. pratensis* L.	23.4-30.0	24.1-28.5	2.2-2.8	2.4-4.0	4.8-6.0	3(4)	16-56, 60, 64
	21. *C. scutata* Thunb.	22.5-27.8	23.9-25.0	1.8-2.6	1.4-2.2	4.0-5.0	3	32
	22. *C. tenera* S.G. Gmel. ex C.A. Mey	24.5-25.0		22.5-27.8	2.0-2.8	2.5-4.0	6.7-8.5	3 16
	23. *C. uliginosa* Bieb.	21.5-27.4	24.0-25.5	2.0-2.7	2.0-3.5	4.8-6.2	3	16
	24. *C. umbellata* Greene	24.7-29.3	26.9-27.3	2.0-2.8	2.8-4.0	5.0-6.5	3	32, 48
	25. *C. wiedemanniana* Boiss.	16.7-22.5	18.4-19.7	1.4-2.0	1.2-2.0	4.0-5.5	3	-
5. Sect. *Cardaminella* Prantl	26. *C. bellidifolia* L.	18.3-22.7	20.7-21.8	1.2-1.8	1.0-2.0	3.5-4.5	3	16
	27. *C. conferta* Jurtz.	21.4-27.8	23.4-25.4	1.6-2.0	1.8-2.5	4.5-6.5	3	48
	28. *C. digitata* Richards.	21.2-26.7	22.6-24.5	1.6-2.0	1.4-2.5	4.0-6.5	3(4)	28, 32, 40, 42
	29. *C. microphylla* Adams	22.0-31.2	23.5-29.2	1.5-2.8	1.2-2.4	4.5-6.0	3(4)	28,32,42,52,64,100
	30. *C. pedata* Regel et Til.	20.0-25.4	22.2-22.8	1.8-2.2	1.4-2.2	4.0-5.0	3	30
	31. *C. purpurea* Cham. et Schlecht.	26.5-31.7	28.2-29.3	2.4-3.0	1.8-2.5	5.0-6.0	3(4)	80, 96
	32. *C. victoris* Busch	21.2-26.7	23.2-25.2	2.0-2.7	2.0-3.8	4.5-5.5	3	28

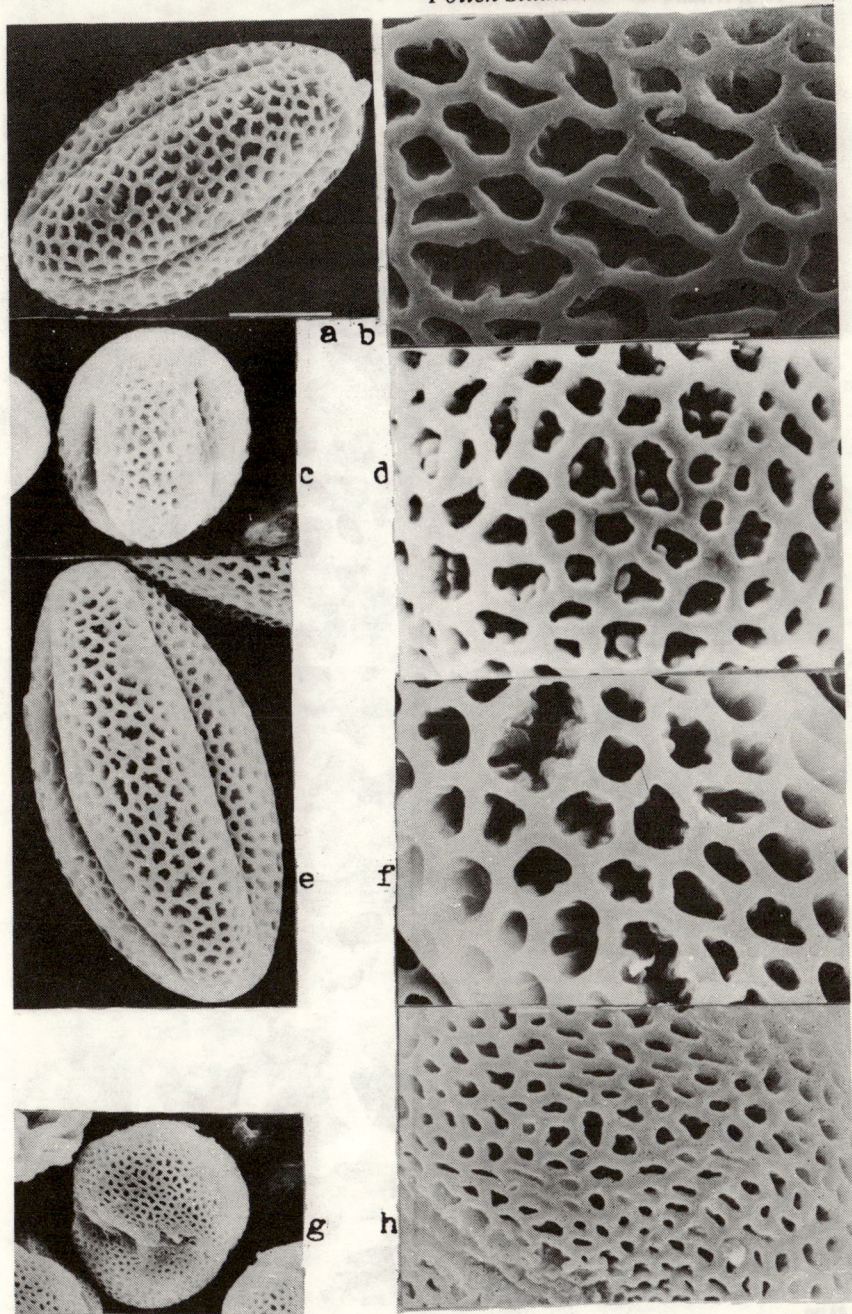

10 μm   2·5 μm

Fig. 1: Pollen grains of *Cardamine* species: a, b—*C. trifida*; c, d—*C. leucantha*, e, f—*C. prorepens*; g, h—*C. fauriei*.

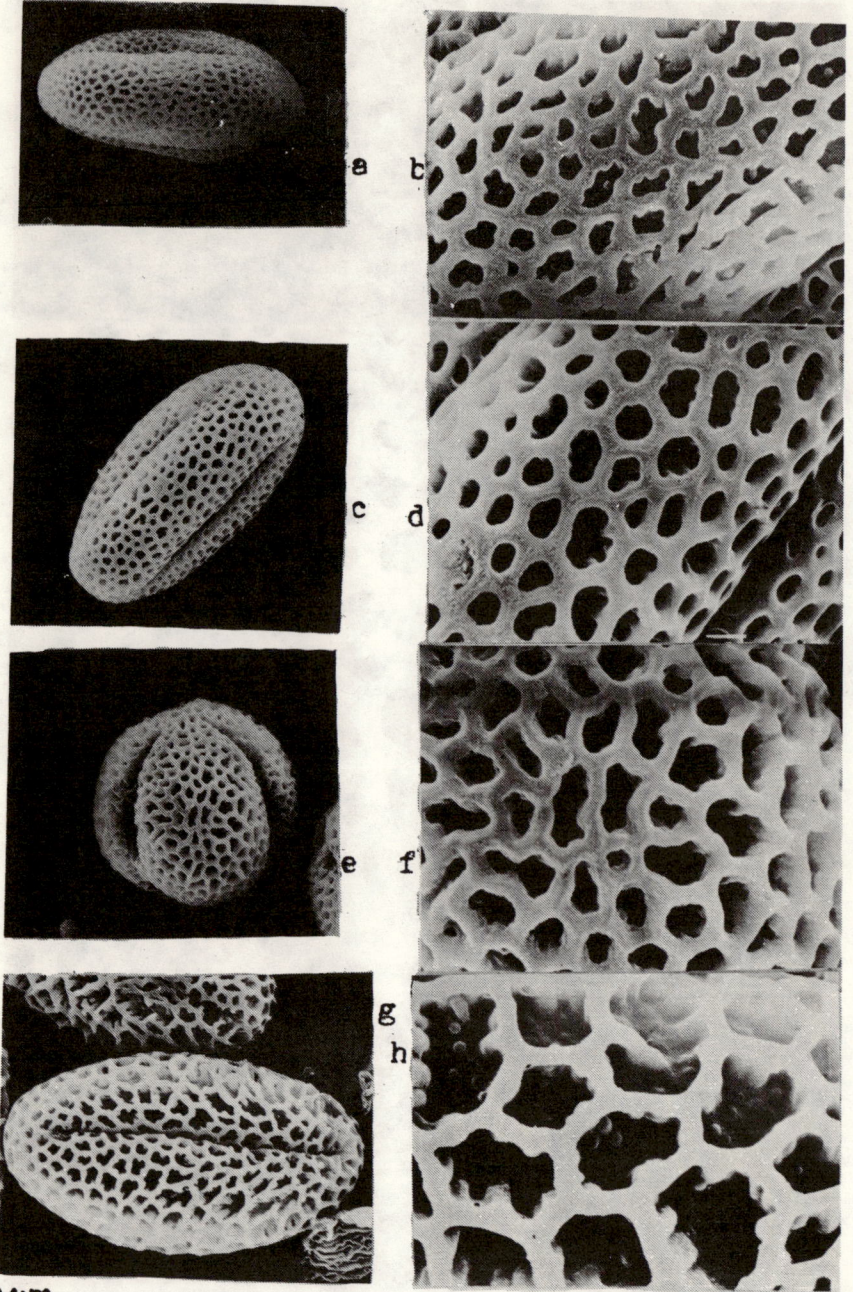

10μm   Fig. 2: Pollen grains of *Cardamine* species: a, b—*C. impatiens*,
c, d—*C. amara*, e, f—*C. scutata*; g, h—*C. umbellata.*   2·5μm

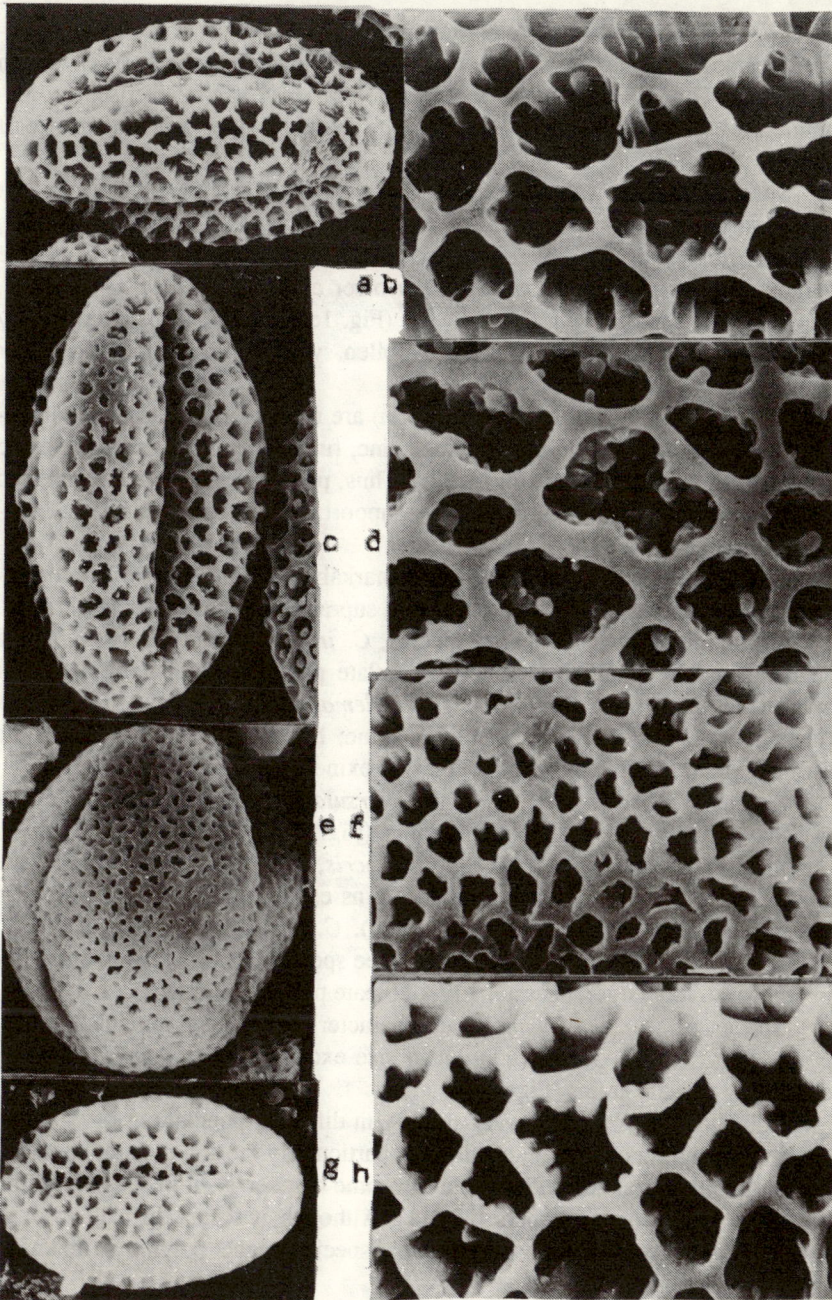

10μm   2·5μm

Fig. 3: Pollen grains of *Cardamine* species; a, b——*C. tenera*; c, d——*C. dentata*, e, f——*C. microphyllia*; g, h——*C. victoris.*

These groups do not correlate taxonomically at the sectional level. However, species or groups of species can be distinguished on the basis of pollen characters. Thus the monotypic sect. *Sphaeroterrhiza* is characterised by medium-sized, coarsely reticulate pollen grains showing wide variations in size of the exine lumina (Fig. 1 a,b).

Species of the Sect. *Macrophyllum* have finely reticulate pollen grains but show variations in other characters. Thus *C. leucantha* (Fig. 1c, d), *C. schinziana* and *C. tementela* are characterised by small-sized, 3-colpate pollen grains with thinner exine and narrower colpi but *C. amaraeformis, C. densiflora, C. macrophylla, C. prorepens* (Fig. 1e, f) and *C. yezoensis* differ by medium-sized, occasionally 4-colpate pollen, with thicker exine and wider colpi.

C. fauriei and *C. lyrate* (sect. *Lyratum*) are readily distinguished by small-sized pollen grains having the thinnest exine, finer reticulation (smallest exine lumina) and narrower colpi (Fig. 1g, h). Thus, pollen characters supplemented by other morpho-anatomical characters support their placement under a separate section [13].

Species of Sect. *Cardamine* show remarkable variations in pollen characters, which also have been found useful in supraspecific groupings [11, 13, 14, 15]. Thus among closely related species, *C. impatiens* and *C. pectinata* are characterised by small-sized, finely reticulate pollen grains having a thinner exine (Fig. 2a, b). *C. amara* and *C. wiedemanniana* also have small-sized, finely reticulate pollen with a thinner exine; however, they differ from the former two species in comparatively larger exine lumina (Fig. 2c, d).

Among *C. parviflora, C. scutata, C. hirsuta* and *C. umbellata*, the former two have finely reticulate pollen grains (Fig. 2e, f) but they are coarsely reticulate in the latter ones (Fig. 2g, h). *C. acris, C. uliginosa* and *C. tenera* are characterised by almost similar pollen grains except that *C. tenera* differs in larger lumina and wider colpi (Fig. 3a, b). *C. pratensis, C. dentata* and *C. nymanii* show similarity with the above three species in exine texture but differ in comparatively larger, occasionally 4-colpate pollen grains (Fig. 3c, d).

Species of sect. *Cardaminella* are characterised by finely reticulate pollen grains (Fig. 3e, f) showing variation in size except that *C. victoris* differs by coarsely reticulate grains (Fig. 3g, h).

Pollen diameter tends to vary slightly in different populations of the same species depending on growth conditions, particularly the soil moisture content, but is directly correlated with the chromosome numbers— it increases with increasing chromosome numbers [22, 11]. Although it seems difficult to draw sharp limits between diploid and tetraploid species, higher polyploids distinctly have larger pollen grains.

Diploid and polyploid species having single cytotypes are characterised by exclusively 3-colpate grains with a negligible percentage of sterile ones; Polyploid species having two or more cytotypes (*C. macrophylla, C. yezoensis,*

C. pratensis, C. dentata, C. nymanii, C. microphyla, etc.) have widely variable, occasionally 4-colpate pollen grains with a remarkably higher percentage of sterile ones. This type of behaviour indicates the instability of certain cyto-types, which possibly represent hybrid swarms of corresponding taxa. Further-more, the presence of 4-colpate grains seems to be correlated with meiotic instability of PMCs as they are consistent in the case of higher sterility but absent in negligible sterility. The proportion of 4-colpate grains also seems to increase with increasing chromosome numbers as they are frequent in higher polyploids (*C. macrophylla, C. dentata, C. nymanii, C. purpurea,* some popu-lations of *C. microphylla* and others).

No karyological report exists for some of these species but pollen charac-ters indicate that *C. schinziana, C. tomentella, C. fauriel, C. lyrata, C. pecti-nata* and *C. wiedemanniana* are diploids while *C. amaraeformis, C. densiflora* and *C. prerepens* are polyploids having several cytotypes.

LITERATURE CITED

1. Erdtman, C. 1952. *Pollen Morphology and Taxonomy: Angiosperms.* Stockholm.
2. Erdtman, C. 1969. *Handbook of Palynology.* Copenhagen.
3. ZaklinskayaA, E.L. 1953. Opisanic pyltsy i spor nekotorykh vidov rastenii polyarnoi tundry. *Trudy Inst. Geol. Nauk (Geol.,* **142**: 3-44.
4. Jankiavichene, R. 1968. Analiz krestotsvenykh Litovskoi SSR. Avtoref. diss. kand. biol. nauk. Vil'nyus.
5. Kupriyanova, L.A. and L.A. Aleshina. 1972. *Pyltsy i spory rastenii flory evropeiskoi chasti SSSR.* Leningrad, **1**: 1-171.
6. Stork, A.L. 1972. Studies in the Aegean flora, XX. Biosystematics of *Malcolmia mari-tima* complex. Opera Bot., **33**: 1-118.
7. Chiguriaeva, A.A. 1973. Morfologiya pyltsy semeistva Cruciferae. In: *Morfologiya pyltsy i spor sovremenykh rastenii.* Leningrad, pp. 93-98.
8. Chiguriaeva, A.A. 1975. Morfologiya pyltsy Brassicaceae. Vopr. Bot. Yugo-Vost., 1:76-119.
9. Chiguriaeva, A.A. 1976. Morfologiya pyltsy Brassicaceae (Thelypodieae i Sohizopetal-eae). Vopr. Bot. Yugo-Vost., **2**: 107-125.
10. Rodman, J.E. 1974. Systematics and evolution of the genus *Cakile* (Cruciferae). Contr. Gray Herb., **205**: 1-246.
11. Spasskaya, N.A. 1979. Kiticheskii obzor vidov roda *Cardamine* L. (Cruciferae JUSS.) evropeiskoi chasti SSSR. Avtoref. diss. kand. biol. nauk. Leningrad State Univ.
12. Jonsell, B. 1986. A monograph of *Farsetia* (Cruciferae). *Symb. Bot. Upsal.,* 25(3): 1-107.
13. Khatri, K.S. 1987. Kriticheskii obzor vidov roda Cardamine L. (Cruciferae JUSS.) ezia-skoi chasti SSSR, i Kaikaja. Avtoref Avtoref. diss. kand. biol. nauk. Leningrad State Univ.
14. Khatri, K.S. 1988. Studies in the Caucasian species of *Cardamine* L. (Cruciferae). *Phyton (Austria)* 28(1): 50-85.
15. Khatri, K.S. 1989. *Cardamine* L. sect. *Cardamine* (Cruciferae) in the Asian territory of Soviet Union. A morphological, anatomical and phylogenetical study with taxonomic re-considerations. *Feddes Report.,* 100(3-4): 80-96.
16. Avetisian, V.E. 1976. Nekotorye modifikatsii sistemy semeistva Brassicaceae. *Bot. Zhurn.,* 61(9): 1198-1203.

17. Schulz, O.E. 1903. Monographic der Gattung *Cardamine. Bot. Jahrb.*, **32**: 280-623.
18. Bolkhovskikh, E.V. et. al. 1969. In: *Khromosomnye chisla tsvetkovykh rastenii.* edited by A.A. Federov, Leningrad.
19. Löve, A. and D. Löve. 1975. *Cytotaxonomical Atlas of the Arctic Flora.* Vaduz.
20. Krogulevich, R.E. and T.S. Rostovtseva. 1984. *Khromosomnye chisla tsvetkovykh rastenii Sibiri i Dalnega Vestoka.* Nevosibirsk.
21. Nishikawa, T. 1986. Chromosome counts of flowering plants of Hekkaide (Japan). *Journ. Hekk. Univ. Educ. Sect.* II B, 37(1): 5-18.
22. Lövkvist, B. 1956. *The Cardamine pratensis* complex. Outline of its cytogenetics and taxonomy. *Symb. Bot. psal.*, 14(2): 1-131.
23. Ellis, R.O. and B.M.G. Jones. 1970. Cardamine - pollen. Watsonia, 8(1):45.
24. Avetisian, E.M. 1950. Uproschenyi metod obrabetki pyltsy. *Bot. Zhurn.*, 35(4): 385-386.